Superjunction Devices

Akshay K

Superjunction Devices

First Edition February 2024

Written by Akshay K

TABLE OF CONTENTS

Page

LIST OF TABLES

LIST OF FIGURES

width at $y = L/2$.

length along the cut-line YY^l of Fig. 5.1 for different charge imbalance factors, k_{eff}. Pillar parameters are $L = 20 \ \mu m$, $W_n = 0.56 \ \mu m$, and $N_d = 8 \times 10^{16} \ cm^{-3}$.

LIST OF ABBREVIATIONS

1-D	1- Dimensional
2-D	2- Dimensional
3-D	3- Dimensional
4H-SiC	4- Hexagonal Silicon Carbide
AC	Alternating Current
BFOM	Baliga's Figure of Merit
BJT	Bipolar Junction Transistors
CSSJ	Charge Sheet Super Junction
DC	Direct Current
FOM	Figure of Merit
GaN	Gallium Nitride
MOSFET	Metal Oxide Semiconductor Field Effect Transistors
NPT	Non Punch Through
PT	Punch Through
Si	Silicon
SiC	Silicon Carbide
SJ	Super Junction
TCAD	Technology Computer Aided Design

LIST OF SYMBOLS

q	Electron charge
V	Voltage
I	Current
R_{ONSP}	Specific on resistance
V_{BR}	Avalanche breakdown voltage
E_g	Band gap of the material
μ_n	Electron mobility
ε	Dielectric constant
V_t	Thermal voltage
n_i	Intrinsic carrier concentration
E_C	Critical electric field of the material
L	Pillar length of a superjunction
L_{opt}	Optimum pillar length of a superjunction
N	Pillar doping of a balanced superjunction
N_{opt}	Optimum pillar doping of a balanced superjunction
N_d	Donor doping concentration
N_a	Acceptor doping concentration
W	Pillar half-width of a balanced superjunction
W_{opt}	Optimum pillar half-width of a balanced superjunction
W_n	n- pillar half-width of a superjunction
W_p	p- pillar half-width of a superjunction
W_d	Depletion width
W_{d0}	Depletion width at zero bias
N_I	Negative interface charge concentration
W_I	Insulator thickness
T	Temperature
T_{dep}	Deposition Temperature
V_R	Reverse bias voltage
t	Drift layer thickness of a 1-dimensional p-n junction

E	Electric field
W_0	Optimum pillar half-width in a balanced superjunction
r	Pillar aspect ratio in a superjunction
r_0	Optimum pillar aspect ratio in a superjunction
r_I	Aspect ratio of the insulator trench
α_n	Impact ionization coefficient of electrons
α_p	Impact ionization coefficient of holes
α_{eff}	Effective impact ionization coefficient of electrons and holes
k_N	Charge imbalance due to pillar doping
k_W	Charge imbalance due to pillar width
k_{eff}	Effective charge imbalance

CHAPTER 1

INTRODUCTION

Devices for generation, distribution and regulation of electric power are important parts of appliances for healthcare, comfort, defence and transportation. Two classes of such devices discussed in this thesis are diodes and MOSFETs. These are used in rectification, amplification and switching applications, and are realized in semiconductor materials. The power level at which these devices operate is decided by the application. Depending on the voltage and current rating, this can vary from 1 Giga Watt at the power station to few Watts for mobile chargers as shown in Fig. 1.1. To cater to this wide range of power levels, power converter circuits are often employed. Depending on the application, these converters could be AC-DC, DC-AC, DC-DC or AC-AC. Although the functionality of these circuits are different, all of them invariably requires power switches. The performance of these switches decide the conversion efficiency of the converter.

The aggregated technical and commercial (ATC) power loss in India is ~ 20 %. This includes power dissipated in semiconductor switches among other factors. Reducing this even by 1 %

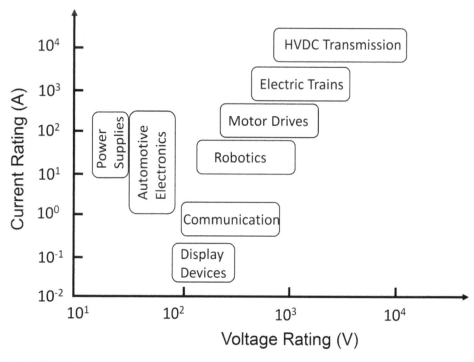

Fig. 1.1 The voltage and current ratings of a wide range of power applications [1].

1

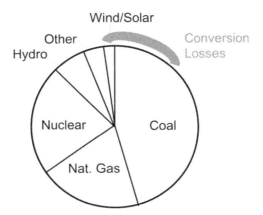

Fig. 1.2 The contribution of various sources of energy to the net generated energy in US. The conversion loss is also indicated. Source: EIA Reports 2009.

could save a few billions of Rupees from the annual budget. Moreover, the government of India has launched the Indian Semiconductor Mission (ISM) on 16th December 2021 in an attempt to achieve self-sufficiency in semiconductor manufacturing for diverse applications; this include devices used in power distribution networks, e-vehicles and defense applications.

The US energy data (see Fig. 1.2) reveals that the conversion loss, mainly due to power dissipation in semiconductor switches, is more than the power generated using all renewable resources together which includes hydro, solar and wind energy. With the ever increasing demand for power caused by $\approx$ 1.1 % annual global population growth and the advent of advanced power hungry applications, there is an immediate need to find ways to reduce power losses in converters.

These factors serve utmost motivation to improve the performance of the switches employed in converters. The requirements from a power switch or device that are usually considered for their design are listed below.

a) High breakdown voltage: During the off state, a reverse bias voltage appears across the device. Depending on the application, this voltage can be between few Volts to tens of kilo Volts. The device must be able to withstand this voltage while allowing only an acceptably low leakage current.

b) Low specific on-resistance: The lower the specific on-resistance of the device, larger the forward current it allows for a given on state voltage drop and chip area. For bipolar devices, it is desirable to have longer carrier lifetime for lower on state voltage drop.

2

c) Fast switching: Faster the devices can switch, lesser the footprint of the associated system due to the smaller value and size of the capacitance and inductor components.

d) High ruggedness: Devices must be capable of safely withstanding harsh operating conditions. This includes circuit conditions such as short circuit as well as ambient conditions such as high temperature and radiation level.

e) Normally-off operation: For a three terminal device, it is desirable to have no current conduction, i.e. to be in off-state, when the control terminal is grounded. This reduces the complexities involved in the design of control circuitry.

Apart from these, high yield and low failure rate in standardization tests are also critical for the commercial manufacturing of power devices. At present, power devices are being realized in silicon as well as wide bandgap materials.

1.1. LIMITATIONS OF SILICON IN POWER DEVICES

So far, power devices fabricated in Si (silicon) have been the workhorse. It is of interest to discuss the performance of a Si switch in terms of the five characteristics listed above. Silicon switches are mostly normally off and thereby avoid complex driving circuits. Apart from this, Si switches underperform in terms of the other four characteristics. Due to the low bandgap and consequently large intrinsic carrier concentration, Si devices cannot reliably operate at temperatures above 100 °C as leakage currents rapidly increase and causes huge power loss. Hence, for reliable operation, Si switches often demand a bulky cooling system that takes away the heat and prevents the device temperature from rising. The minimum achievable specific on-resistance (R_{ONSP}) for a given breakdown voltage (V_{BR}) for a conventional silicon diode (see Fig. 1.3(a)) is referred to as *silicon limit*. A quantitative measure of this limit is given by [1]:

$$R_{ONSP} = \frac{4V_{BR}^{2}}{\varepsilon_s \mu_n E_C^{3}}$$

(1.1)

where ε_s is the semiconductor permittivity, μ_n is the electron mobility, E_C is the critical field for avalanche breakdown and $\varepsilon_s \mu_n E_C^3$ is popularly known as the Baliga's figure of merit (BFOM). Higher the BFOM, lower the R_{ONSP} achievable for a given V_{BR}. Due to the low bandgap (E_g) property of Si, the E_C is low (~ 300 kV/cm) which results in a high R_{ONSP} for a given V_{BR}. This has limited the performance of unipolar Si power devices.

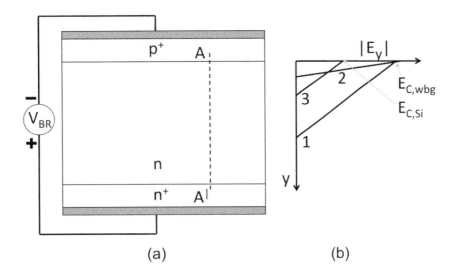

(a)   (b)

Fig. 1.3 (a) A conventional p^+n junction reverse biased near breakdown. (b) The breakdown field distribution along AA$^|$ for the device in (a). The lines marked 1 and 2 correspond to wide bandgap materials (SiC or GaN) while 3 is a Si device. The area under lines 2 and 3 is same.

Table 1.1: Material parameters of Si, GaN and 4H-SiC

Material	Bandgap (eV)	Critical Electric Field (MV/cm)	Electron Mobility (cm^2/Vs)	Thermal Conductivity (W/mK)
Si	1.12	0.3	1300	130
GaN	3.45	3-3.5	2000	110
4H-SiC	3.26	2.5-3	700-900	700

1.2 WIDE BANDGAP MATERIALS IN POWER DEVICES

To improve the performance of power device, there was an effort to move beyond Si towards wide bandgap semiconductors, namely SiC and GaN. The motive behind this can be easily understood from Eq. (1.1) as the R_{ONSP} for a given V_{BR} is inversely related to E_C^3. The value of E_C and a few other important material parameters of Si, SiC and GaN are given in Table 1.1.

The following discussion illustrates qualitatively how the wide bandgap materials improve the V_{BR}-R_{ONSP} trade-off. Consider the sketch of the breakdown electric field along the cutline, AA$^|$

(see Fig. 1.3(a)), given in Fig. 1.3(b) for a low bandgap material (Si) and a wide bandgap material (SiC/GaN). Curves 1 and 2 correspond to a wide bandgap material based device while 3 corresponds to a Si device. Curves 1 and 3 correspond to the case of equal doping concentration of the drift region evident from their equal slope. This would mean that for a given doping level, wide bandgap devices can have a much larger V_{BR}. Curves 2 and 3 have equal area under its distributions and hence the same V_{BR}. In this case, curve 2 has a much larger slope than curve 3 which implies a larger doping and consequently a lower R_{ONSP}.

The GaN devices offer higher V_{BR} than SiC counterparts theoretically due to the higher bandgap. However, the high trap density in GaN grown on SiC, Si or Sapphire substrate has hindered the realization of GaN vertical devices. Vertical architecture is a key factor to increase the current density in power devices. The technology for homoepitaxial GaN growth to make GaN on GaN wafers is seen as a ray of hope to push GaN vertical devices and compete with SiC counterparts. Hence, GaN vertical devices are a strong academic research topic currently.

However, SiC devices are easier to make using the existing Si technology. The advantages of SiC over GaN are manifold. Firstly, it has the same native oxide as Si making oxidation, masking etc. convenient. Secondly, vertical devices can be realized easily. Further, the bulk trap density of SiC is lesser than GaN devices. This would mean a lower leakage current and higher reliability. Further, SiC devices can operate at high temperatures up to 400 °C when packaged properly. Hence, much of the efforts are to replace Si devices with its SiC counterparts [2].

1.3. SUPERJUNCTION

1.3.1. Structure

The quadratic dependence of R_{ONSP} on V_{BR} given by Eq. (1.1) can be changed to linear if the n-pillar of a conventional diode is replaced with a stack of alternating n- and p-pillars resulting in a superjunction. The typical structure of a superjunction diode is given in Fig. 1.4(a) while that of a superjunction vertical power MOSFET is given in Fig. 1.5; the results discussed in this work are applicable for the drift layer design of both of these devices. A cross-sectional scanning capacitance microscopy (SCM) image of a fabricated superjunction is given in Fig. 1.4(c) [3]. A balanced superjunction is an ideal device having equal n- and p- pillar charge, i.e.

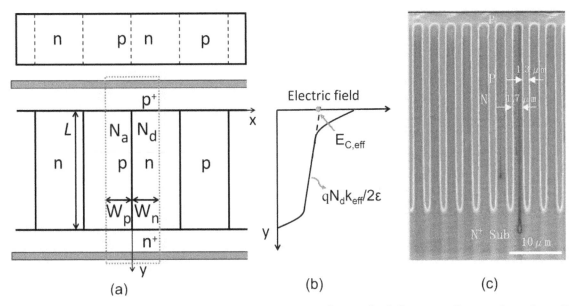

Fig. 1.4 (a) Horizontal and vertical sectional view of a typical linear cell superjunction diode with multiple pillars. Box of dotted lines represents a unit cell of the device. (b) Breakdown electric field distribution along $x = W_n$ of the device in (a). (c) Cross-sectional SCM image of a fabricated superjunction device (reproduced from [3]).

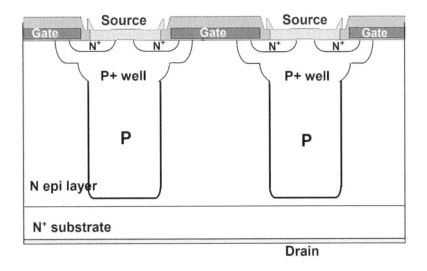

Fig. 1.5 Typical balanced superjunction MOSFET structure with multiple pillars.

$N_d = N_a = N$ and $W_n = W_p = W$. The relation between R_{ONSP} and V_{BR} for a balanced superjunction is given by

$$R_{ONSP} = 8W \times \frac{V_{BR}}{\varepsilon_s \mu_n E_C{}^2}$$ (1.2)

6

TABLE 1.2

Values of constants in Eq. (1.3)-(1.4)

Constant	Material	
	Si [14],[15]	4H-SiC [16],[17]
A_n, A_p (cm^{-1})	9.0×10^5	1.14×10^9, 6.85×10^6
B_n, B_p (V cm^{-1})	1.8×10^6	3.8×10^7, 1.41×10^7
μ_{max} (cm^{2}V^{-1}s^{-1})	1417	950
μ_{min} (cm^{2}V^{-1}s^{-1})	52	40
N_0 ($\times 10^{16}$ cm^{-3})	9.68	20.00
γ	0.68	0.76

The invention of superjunction devices dates back to 1978 [4]. Later, it was introduced in the patent field in the 1980s and 1990s [5]–[10], but its commercialization took place only in the late 1990s [11] with two trademark products namely, CoolMOS [11], [12] and MDMesh [13], from Infineon and ST Microelectronics respectively.

The parameters involved in the design of a balanced superjunction are pillar length, L, doping, N, and width, $2W$. The lateral p-n junction has a zero bias depletion width, $2W_d$.

1.3.2. Simulation

Technology Computer Aided Design (TCAD) simulations play an unavoidable role in the modern semiconductor sector. In this book, we use well calibrated TCAD simulations for the analysis and design of superjunctions and also for validation of the analytical models derived in the later chapters. Our simulations employ the SELB impact ionization model given by

$$\alpha_{n,p} = A_{n,p} \exp\left(-B_{n,p} / E_y\right), \tag{1.3}$$

during reverse bias and ANALYTIC CONMOB concentration dependent mobility model given by

$$\mu_n = \mu_{min} + \frac{\mu_{max} - \mu_{min}}{1 + \left(N_d / N_0\right)^\gamma}, \tag{1.4}$$

during forward bias. Here, α_n (α_p) is the impact ionization coefficient of electron (hole) and μ_n is the electron mobility. The fitting parameters of these models for Si and 4H-SiC materials have

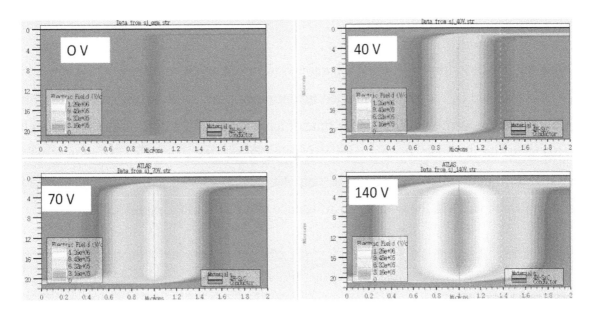

Fig. 1.6 TCAD simulated electric field contours in a 4H-SiC superjunction device having $V_{BR} =$ 3.6 kV at different applied reverse bias voltages, $V_R = 0, 40, 70, 140$ V. The vertical PN junction depletes the pillars at ≈ 140 V. $L = 20$ μm, $W = 1$ μm, and $N = 8 \times 10^{16}$ cm^{-3}.

been calibrated against measured data [14]-[17] and are listed in Table 1.2. Further, the simulations predict the measured V_{BR} of a Si superjunction [3] with < 5 % error.

Simulations using the well calibrated TCAD set up are used to analyze the off- and on- state operation of a superjunction and is discussed below.

1.3.3. Operation

1.3.3.1 Off State

When the reverse bias voltage across the device is increased in small steps, the pillars get laterally depleted at a low voltage as it is designed to have $W << L$ (see Fig. 1.6). Once the pillars are depleted, for further increment in reverse bias, the field lines from the n$^+$ travel the entire pillar length, as it do not find any charge in the pillar to terminate on. As a result, the electric field is uniform along L and $\approx E_C$ as shown in Fig. 1.4(b), unlike a conventional diode where the field is triangular. Hence, the V_{BR} of an SJ can be higher compared to that of a conventional junction. The V_{BR} of an SJ can be estimated as the area under its breakdown field distribution. From Fig. 1.4(b), we get

$$V_{BR} = L \times E_C \tag{1.5}$$

1.3.3.2 On State

When the device is forward biased, a current flows from anode to cathode. This current is the sum of electron and hole current conducted predominantly by the n- and p- pillars respectively. However, as electron mobility is $\approx 2-3$ times higher than hole mobility, the electron current component would dominate the net current flowing across the electrodes. Moreover, when SJ structures are employed as drift layer in unipolar devices, the p-pillars do not conduct any current. For these reasons, the specific on-resistance of an SJ is often expressed as,

$$R_{ONSP} = \frac{L}{qN_d\mu_n}\left(\frac{2W}{W-W_d}\right).$$

(1.6)

1.3.4 Silicon Superjunction Fabrication

The methods by which Si SJs are commercially manufactured can be broadly categorized into two: multiepitaxy and trench. The mutiepitaxy uses a number of masked implantations and epitaxial layer growth (see Fig. 1.7). The key advantage of this method is its capability to adjust the doping levels of p- and n- regions in each layer. However, the large number of process steps, especially for high V_{BR} devices that demand longer drift region, makes this method expensive, time consuming and prone to severe process variations.

In contrast, trench technology is based on formation of deep trenches and subsequent side-wall epitaxial filling to form the deep p- and n-column structures (see Fig. 1.8). The advantage of the trench technology lies in the smooth shape of the p/n junction. However, precise process control is mandatory to fill every trench perfectly. Hence, it is prone to process variations.

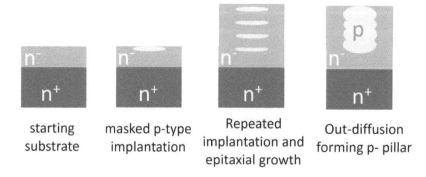

Fig. 1.7 Key steps involved in the fabrication of superjunction using mutiepitaxy method

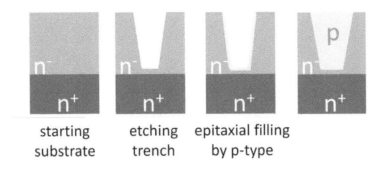

starting | etching | epitaxial filling
substrate | trench | by p-type

Fig. 1.8 Key steps involved in the fabrication of superjunction using trench technology

1.3.5 Wide Bandgap Superjunction Devices

Superjunction concept was originally introduced in the context of silicon devices. Later, it has been extended to other wide bandgap materials like SiC [18]-[20] and GaN [21]-[26] to combine the advantage of linear V_{BR}-R_{ONSP} relation and the large value of E_C.

It is to be noted that although different materials offer different technological challenges to realize a superjunction, the theory of superjunction operation and the analytical design equations are valid across materials upon making changes to the material specific parameters (eg. impact ionization coefficients, mobility etc.). Hence, a model developed to design the optimum superjunction parameters has potentially a large impact. It can help technologists working across materials to steer their efforts more effectively towards realizing the best device for a target application with minimum cost and effort.

1.3.5.1 Fabrication

Having said this, the technological challenges involved in realizing a superjunction are not to be overlooked. The fabrication of a Si superjunction is the easiest compared to other materials due to the mature Si fabrication technology. As a result commercial SJs in Si are available, whereas commercial SJs of no other material are commercially available yet. The most investigated wide bandgap material for SJ is SiC. However, several fabrication challenges are to be solved before they can be manufactured reliably and profitably. For instance, processes that are critical for the fabrication of SJ like trench etching, ion implantation, diffusion, annealing and epitaxial growth are relatively still in infancy for SiC as compared to Si. The fabrication of SJ in 4H-SiC has been problematic. The first two reported works could not make a functional SJ

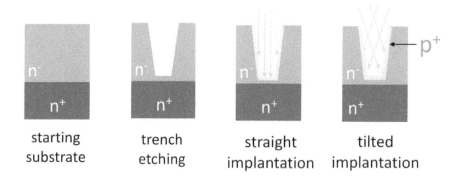

Fig. 1.9 Key steps involved in the fabrication of SiC superjunction [20]

device [18], [19]. The first functional 4H- SiC SJ was reported by Zhong et al. [20], [27] using trench etching and sidewall implantation of p-dopant. However, the fabrication of this 1.35 kV SJ device required six high-energy implantations followed by a high temperature anneal for creating a p^+ liner along the sidewalls of the trench (see Fig. 1.9). Also, the p-dopant activation efficiency varied between $20 - 65$ % in the annealing temperature range, and required a trial and error approach to locate the temperature corresponding to the "optimum charge balance" that yields the maximum V_{BR}. Considering that the charge imbalance level in Si SJ can go up to 20 % [28], such imbalance levels could be even higher in 4H-SiC SJ due to this added difficulty in controlling the p-dopant activation efficiency [17]. In addition, the above steps of fabricating p-pillars in a 4H-SiC SJ are tedious, expensive and prone to causing severe wafer damage. Hence, alternative solutions for realizing the p- pillars in 4H-SiC SJs are highly sought after.

Based on the above sections, we conclude that the analytical design of SJ is an important research problem due to the following reasons. First, it quickly gives deep physical insights into the optimal device design that are often difficult to obtain from a tedious trial and error based simulation study. Second, it can augment the data-driven approaches as follows. Data-driven approaches lack physical insights and consequently could lead to misinterpretation in the presence of poor quality data. As analytical designs are derived by considering physical effects, their accuracy is independent of the quality of the data at hand and hence the insights derived from it can help to avoid possible misinterpretations. Third, analytical solutions can be expressed in a material independent form which makes it useful for the design of SJ for a wide range of materials.

1.6 ORGANIZATION OF THE BOOK

In Chapter 2, we critically review the prior solutions to design an SJ. Chapters 3 and 4 give our improved solutions for an ideal balanced SJ and a practical imbalanced SJ, respectively. In Chapter 5, the practicability, modeling and design of a novel device, namely Charge Sheet SJ in 4H-SiC material are presented. Chapter 6 gives the summary and conclusions of our work.

1.7 REFERENCES

[1] B. J. Baliga, Fundamentals of Power Semiconductor Devices, Springer Publishers, 2008.

[2] B. J. Baliga, Silicon Carbide Power Devices, World scientific, 2006.

[3] J. Sakakibara, Y. Noda, T. Shibata, S. Nogami, T. Yamaoka and H. Yamaguchi, "600 V-class super junction MOSFET with high aspect ratio P/N columns structure", *Proc. 20th ISPSD*, pp. 299-302, May 2008, doi: 10.1109/ISPSD.2008.4538958.

[4] S. Shirota and S. Kaneda, "New type of varactor diode consisting of multilayer p-n junctions," J. Appl. Phys., vol. 49, no. 12, pp. 6012–6019, 1978.

[5] D. J. Coe, "High voltage semiconductor devices," European Patent 0 053 854, Jun. 16, 1982.

[6] D. J. Coe, "High voltage semiconductor device," U.S. Patent 4 754 310, Jun. 28, 1988.

[7] X. Chen, "Semiconductor power devices with alternating conductivity type high-voltage breakdown regions," U.S. Patent 5 216 275, Jun. 1, 1993.

[8] J. Tihanyi, "Power MOSFET," U.S. Patent 5 438 215, Aug. 1, 1995.

[9] T. Fujihira et al., "Semiconductor device with alternating conductivity type layer and method of manufacturing the same," U.S. Patent 6 1683 347 B1, Jul. 7, 1997.

[10] F. Udrea, "Semiconductor Device," U.S. Patent 6 111 289, Apr. 12, 1999.

[11] G. Deboy, M. März, J.-P. Stengl, H. Strack, J. Tihanyi, and H. Weber, "A new generation of high voltage MOSFETs breaks the limit line of silicon," in IEDM Tech Dig., pp. 683–685, Dec. 1998.

[12] L. Lorenz, G. Deboy, A. Knapp, and M. Marz, "COOLMOS—A new milestone in high voltage power MOS," in Proc. 11th Int. Symp. Power Semiconductor Device ICs, 1999, pp. 3–10.

[13] M. Saggio, D. Fagone, and S. Musumeci, "MDmesh: Innovative technology for high voltage power MOSFETs," in Proc. 12th Int. Symp. Power Semiconductor Device ICs, 2000, pp. 65 68.

[14] W. Fulop, "Calculation of avalanche breakdown voltages of silicon pn junctions," *Solid-St. Electron.*, vol. 10, no. 1, pp. 39-43, 1967, doi: 10.1016/0038-1101(67)90111-6.

[15] B. Van Zeghbroeck, Principles of Semiconductor Devices. Englewood Cliffs, NJ, USA: Prentice-Hall, Dec. 2009.

[16] M. Roschke and F. Schwierz, "Electron mobility models for 4H, 6H, and 3C SiC," *IEEE Trans. Electron Devices*, vol. 48, no. 7, pp. 1442–1447, Jul. 2001, doi: 10.1109/16.930664.

[17] L. Yu and K. Sheng. "Modeling and optimal device design for 4H-SiC super-junction devices," *IEEE Trans. Electron Devices,* vol. 55, no. 8, pp. 1961-1969, Jul. 2008, doi: 10.1109/TED.2008.926648.

[18] R. Kosugi, Y. Sakuma, K. Kojima, S. Itoh, A. Nagata, T. Yatsuo, Y. Tanaka and H. Okumura, "Development of SiC super-junction (SJ) device by deep trench- filling epitaxial growth," Mater. Sci. Forum, vols. 740–742, pp. 785–788, 2013.

[19] R. Kosugi, Y. Sakuma, K. Kojima, S. Itoh, A. Nagata, T. Yatsuo, Y. Tanaka and H. Okumura, "First experimental demonstration of SiC superjunction (SJ) structure by multi-epitaxial growth method," in Proc. Int. Symp. Power Semiconductor Device ICs, pp. 346–349, June 2014.

[20] X. Zhong, B. Wang, and K. Sheng, "Design and experimental demonstration of 1.35 kV SiC super junction Schottky diode," in Proc. Int. Symp. Power Semiconductor Device ICs, pp. 231–234, June 2016.

[21] H. Ishida et al., "GaN-based natural super junction diodes with multichannel structures," in Proc. Int. Electron Devices Meeting, 2008, pp. 1–4.

[22] Nakajima, Y. Sumida, M. H. Dhyani, H. Kawai, and E. M. Narayanan, "GaN-based super heterojunction field effect transistors using the polarization junction concept," IEEE Electron Device Lett., vol. 32, no. 4, pp. 542–544, Apr. 2011.

[23] Z. Li, and T. P. Chow, "Design and simulation of 5–20-kV GaN enhancement-mode vertical superjunction HEMT," IEEE Trans. Electron Devices, vol. 60, no. 10, pp. 3230-3237, 2013.

[24] V. Unni et al., "2.4kV GaN polarization superjunction Schottky barrier diodes on semi-insulating 6H-SiC substrate," in Proc. 26th Int. Symp. Power Semiconductor Device ICs, 2014, pp. 245–248.

[25] Song et.al., "Design and optimization of GaN lateral polarization-doped super-junction (LPSJ): An analytical study " in Proc. Int. Symp. Power Semiconductor Device ICs, 2015, pp. 125-128.

[26] H. Hahn et al., "Charge balancing in GaN-based 2-D electron gas devices employing an additional 2-D hole gas and its influence on dynamic behaviour of GaN-based heterostructure field effect transistors," J. Appl. Phys., vol. 117, no. 10, pp. 104508-1–104508-7, 2015.

[27] X. Zhong, B. Wang, J. Wang and K. Sheng, "Experimental Demonstration and Analysis of a 1.35-kV 0.92-mΩ.cm^2 SiC Superjunction Schottky Diode", *IEEE Trans. Electron Devices,* vol. 65, no.4, pp.1458-1465, 2018, doi: 10.1109/TED.2018.2809475.

[28] E. Napoli, H. Wang, and F. Udrea, "The effect of charge imbalance on superjunction power devices: An exact analytical solution", *IEEE Electron Device Lett.,* vol. 29, no. 3, pp. 249–251, Mar. 2008, doi: 10.1109/LED.2007. 915375.

CHAPTER 2

REVIEW

The design of the drift layer of power devices involves deciding its doping and dimensions for meeting a target V_{BR} and correspondingly the least R_{ONSP}. This is critical in power device design because lower the R_{ONSP}, lower the power dissipation and cheaper and lighter the cooling requirements. In this chapter, we review the prior solutions available for the design of two drift layer architectures, namely - 1-dimensional (1-D) junctions and superjunctions (SJs), and point out the limitations of the latter. SJs offer lower R_{ONSP} for a given V_{BR} than 1-D junctions as the two dimensional electric field distribution allows the former to have a higher doping than latter. Several works that treated the SJ design problem assumed the n- and p- pillar charges to be equal to simplify the analyses; this structure corresponds to an ideal device namely balanced SJ, which yields the lower limit of R_{ONSP} achievable for a target V_{BR}. However, it was realized that the process variations during fabrication inevitably introduce some degree of charge imbalance and hence need to be accounted for real device design.

In a power device, the electric field at the edges becomes larger than at the middle of the device, leading to a reduction of the breakdown voltage. Various edge terminations have been proposed to mitigate this effect [1]-[4]. Fig. 2.1(a) shows a conventional p-n junction with multiple floating diffused ring edge termination. Super-junction devices need a different type of edge termination since their drift region doping is up to 10 times higher. As an example, Fig. 2.1(b) shows the edge termination employed in a silicon superjunction fabricated using multi-epitaxy method. Here, laterally and vertically graded doping profiles are realized in the termination region using several p-type implants. These doping profiles are optimized so that they effectively mitigate the electric field crowding at the edges. Generally, the design of the edge termination is discussed separately from the design of the middle of the device, where the junction can be treated as plane parallel. In this thesis, we focus on the design of the plane parallel portion of the junction.

In Section 2.1, we review the available design of the drift layer. In Section 2.2 and Section

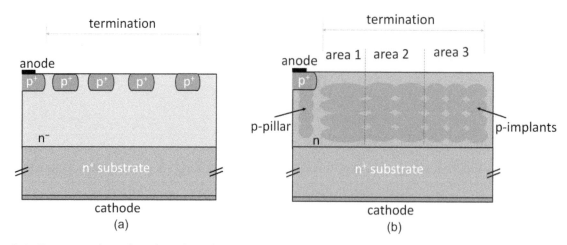

Fig. 2.1 Cross section showing the edge termination structure of a 1-dimensional p-n junction (a) and a superjunction fabricated using multi-epitaxy method (b) [3].

2.3, we review the various available approaches to design ideal balanced and practical imbalanced SJs, respectively, and point out the limitations of each approach. In section 2.5, we summarize the gaps in literature and the objectives of the present work.

2.1 1-DIMENSIONAL P⁺-N JUNCTION DESIGN

The design problem is to find the values of the drift layer doping, N, and thickness, t for a given V_{BR} and minimum R_{ONSP} (Fig. 2.2(a)). The R_{ONSP} of a 1-D junction is given by

$$R_{ONSP} = \frac{L}{qN\mu_n}. \tag{2.1}$$

A simple and therefore widely used solution suggests choosing N and t such that the electric field distribution at breakdown decreases linearly from a peak value, E_C, at the blocking junction to zero at the drift layer–substrate junction; this corresponds to a Non-Punch Through (NPT) structure (see Fig. 2.2(b)) [5]. In this approach, N and t are given by

$$N = \frac{\varepsilon E_C^{\,2}}{2qV_{BR}}, \quad t = 2V_{BR}/E_C \tag{2.2}$$

where E_C is the critical electric field strength of the material and is given by

$$E_C(Si) = 4010 \times N^{1/8} \text{ V/cm}, \quad E_C(4H - SiC) = 3.3 \times 10^4 \times N^{1/8} \text{ V/cm} \tag{2.3}$$

15

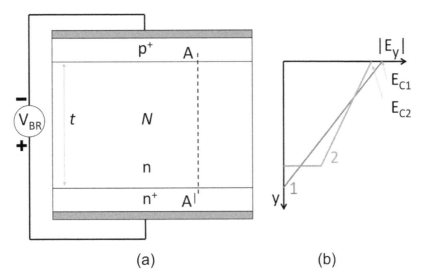

(a)   (b)

Fig. 2.2 (a) Conventional 1-D junction and the (b) breakdown electric field distributions corresponding to the designs in literature; curve 1(2) corresponds to [5] ([6]).

Using (2.2) in (2.1), R_{ONSP} can be expressed as

$$R_{ONSP} = \frac{4V_{BR}^2}{\varepsilon_s \mu_n E_C^3} \qquad (2.4)$$

It is well-known that Eq. (2.4) do not yield the lower limit of R_{ONSP} for a target V_{BR}, often referred to as the unipolar limit. Nevertheless, due to the simplicity of the solution, it is still being used for determining the theoretical unipolar limit of materials [7]-[9] and even for designing practical devices [10]-[11]. However, a mathematically rigorous solution using the method of finding extrema was presented in [6]. In this approach the R_{ONSP} expressed as a function of t alone was differentiated with respect to t and set to zero to obtain the optimum value of t. This yields

$$N = \frac{4\varepsilon E_C^2}{9qV_{BR}} \quad , \ t = 3V_{BR}/2E_C \qquad (2.5)$$

where E_C can be assumed to be same as (2.3). Using (2.5) in (2.1), R_{ONSP} can be expressed as

$$R_{ONSP} = \frac{27V_{BR}^2}{8\varepsilon_s \mu_n E_C^3} \qquad (2.6)$$

It is to be noted that R_{ONSP} given by (2.6) is $\approx 15\ \%$ lower than that yielded by (2.4). Also, the N

and t given by (2.5) is lower than (2.2) suggesting that a punch through (PT) design offers a lower R_{ONSP} as compared to a NPT counterpart for the same V_{BR}. Several numerical investigations on 4H-SiC [12], GaN [13] and diamond [14] Schottky diodes have later confirmed the superiority of PT structures over NPT.

The above approaches assumed a uniform doping profile for the drift region. However, it was shown that a non-uniform doping profile, $N(y)$, given by

$$N(y) = \frac{\varepsilon E_C^2}{3qV_{BR}\sqrt{1 - 2E_C y/3V_{BR}}} \tag{2.7}$$

and t given by (2.4) yields an R_{ONSP},

$$R_{ONSP} = \frac{3V_{BR}^2}{\varepsilon_s \mu_n E_C^3} \tag{2.8}$$

that is 12.5 % lower than that given by (2.6) [15]. Eq. (2.7) was analytically derived by applying the calculus of variation and the method of Lagrange multipliers [16].

2.2 BALANCED SUPERJUNCTION DESIGN

The R_{ONSP} offered by drift layer can be further lowered by replacing the 1-D junction with a superjunction. It is of interest to determine the lower limit of R_{ONSP} for a target V_{BR}; this occurs for a *balanced* superjunction having equal n- and p- pillar charge. The design problem is to find the values of the pillar parameters namely doping, N, length, L, and width, W, for a target V_{BR} and minimum R_{ONSP}.

Before discussing the various reported design methods in detail, a tabular comparison of them is provided in Table 2.1. The factors used for the comparison are a) the type of modeling approach adopted, features of the b) V_{BR} and c) R_{ONSP} models reported and the equations derived to design the optimum parameters, d) L_{opt}, e) W_{opt} and f) N_{opt}. These methods are detailed below.

2.2.1 Fujihira Method

This is the first work to report a model that gives the relation between V_{BR} and R_{ONSP} of a balanced superjunction [17]. Here, the breakdown electric field along the critical path, $x = W$, is approximated as a curve with a portion, $y \leq W$, linearly decreasing from $|E_{y\,max}| = E_C$ with a slope $= qN/\varepsilon$ and the remaining portion as constant, $|\Delta E_{y\,max}|$, with distance (see Fig. 2.3). The "optimum" design criteria that minimizes R_{ONSP} is proposed to be $|E_{y\,max}| = 2*|E_{x\,max}| = 2*|\Delta E_{y\,max}| = E_C$, where $|E_{x\,max}|$ and E_C are given by

$$| E_{x\max} |= \frac{qN}{\varepsilon} W \quad , \quad E_C = 2000 \times N^{1/7}\,\text{V/cm} \tag{2.9}$$

The R_{ONSP}-V_{BR} trade-off in a balanced superjunction was shown to be

$$R_{ONSP} = \frac{8WV_{BR}}{\mu \varepsilon E_C^{\,2}} \tag{2.10}$$

and is plotted in Fig. 2.4 for several values of W.

Limitations: (a) The work do not yield pillar parameters for minimum R_{ONSP}, as shown in later studies. This work can be used to find the V_{BR} and R_{ONSP} of a superjunction with a given set of pillar parameters. However the design problem is the inverse: to determine the L_{opt}, N_{opt} and W_{opt} for a target V_{BR} and minimum R_{ONSP}. (b) Avalanche breakdown happens when the ionization integral is unity along the critical path. However, this work uses a less accurate E_C based approach and further assumes it to be dependent only on pillar doping. However, actual E_C is also geometry dependent. This work uses an oversimplified picture to do the modeling and hence yields results that significantly deviate from TCAD simulations. (c) The zero bias depletion

Table 2.1 Comparison of the prior balanced superjunction design methods

Method	Fujihira [17]	Chen [18], Strollo [19]	Yu [20]
Model Type	Physical	Semi Empirical	
V_{BR} model	E_C based	Ionization integral based	
R_{ONSP} modeling	W_d neglected	W_d and $\mu(N)$ neglected	Includes all effects
Design of L	No equation	$L = \left(a\left(1+b\times f(\frac{W}{L})\right)V_{BR}^{\,7}\right)^{\frac{1}{6}}$	No equation
Design of W	From technology		
Design of N	$\dfrac{qNW}{\varepsilon} = \dfrac{E_C}{2}$	$\dfrac{qN}{2\varepsilon} L^2 = \dfrac{V_{BR}}{f(W/L)}$	No equation

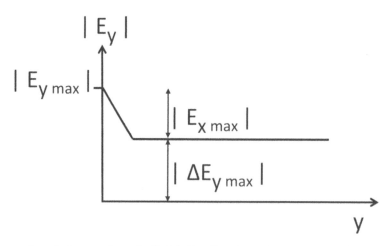

Fig. 2.3 Approximate breakdown electric field distribution along $x = W$ of a balanced SJ.

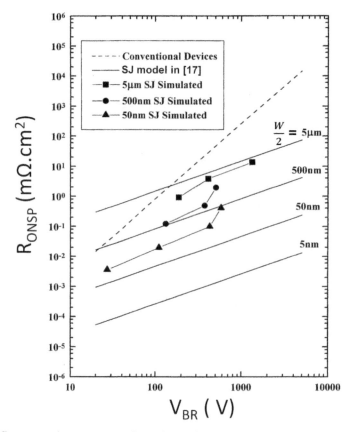

Fig. 2.4 The specific on resistance as a function of breakdown voltage predicted by the model [17] and simulations for a balanced Si superjunction (reproduced from [17]).

width in the n- pillar is neglected. This leads to underestimation of actual R_{ONSP} values, especially for lower values of W and gives the incorrect insight that the R_{ONSP} can be indefinitely reduced by reducing W.

19

2.2.2 Chen Method

This work [18] improved the models reported in Ref.[17] by adopting a more mathematically sound modeling approach. The breakdown electric field distribution along the critical path was modelled as a sum of uniform value and a decaying exponential and given by

$$E(y) = \frac{E_C}{2}\left[1 + \exp\left(\frac{2y/L - 1}{f}\right)\right]$$

(2.11)

where $f = 1.484W/L$. This is used to evaluate the ionization integral,

$$\int 1.8 \times 10^{-35} E^7 dy = 1$$

(2.12)

This results in

$$E_C = 1.24 \times 10^6 V_{BR}^{-1/6}(1 + 21.7f)^{-1/6} \text{ V/cm}$$

(2.13)

Hence, the geometric and doping dependency of E_C has been accounted. The design starts by choosing a target V_{BR} and assuming a value of f. Further, the optimum pillar doping length and width are obtained as follows.

$$N = \frac{\varepsilon E_C}{fqL}$$

(2.14)

$$L = \frac{2V_{BR}}{E_C}$$

(2.15)

$$W_n = \frac{fL}{1.484}$$

(2.16)

Limitations: Although this work improves over the prior work [17], it do not yield the minimum R_{ONSP} for a target V_{BR}. Further, the R_{ONSP} calculations are made by neglecting the doping dependency on mobility and zero bias depletion width of n-pillar which underestimates the actual value of R_{ONSP}.

2.2.3 Strollo Method

This paper [19] reports a model, very similar to that reported in Ref. [18], and demonstrates an improved tradeoff relation between V_{BR} and R_{ONSP} of a balanced superjunction, compared to that presented in Ref. [17]. Solving the 2-D Poisson equation inside the device yields the electric field distribution from which the maximum, E_{max}, and minimum, E_{min}, value of electric field along $x = W$ can be obtained and given by,

$$E_{max} = \frac{V_R}{L} + \frac{V_D}{L} f\left(\frac{W}{L}\right) \quad , \quad E_{min} = \frac{V_R}{L} - \frac{V_D}{L} f\left(\frac{W}{L}\right)$$

(2.17)

where V_R is the reverse bias voltage applied and V_D and $f(t)$ are given by

$$V_D = (qN/2\varepsilon) L^2 \quad , \quad f(t) = 1.697t - 0.7524t^2 - 1.9 \times 10^{-4} t^3$$

(2.18)

A simple electric field function is proposed empirically

$$E(y) = \frac{V_R}{L} + \frac{V_D}{L} f\left(\frac{W}{L}\right) \exp\left(-\frac{2y}{f(W/L)L}\right)$$

(2.19)

Using (2.19) for the evaluation of ionization integral yields

$$V_R^{\,7} + \frac{7V_R^{\,6} V_D}{2} f\left(\frac{W}{L}\right)^2 + \frac{21 V_R^{\,5} V_D^{\,2}}{4} f\left(\frac{W}{L}\right)^3 + \frac{35 V_R^{\,4} V_D^{\,3}}{6} f\left(\frac{W}{L}\right)^4 + \frac{35 V_R^{\,3} V_D^{\,4}}{8} f\left(\frac{W}{L}\right)^5$$
$$+ \frac{21 V_R^{\,2} V_D^{\,5}}{10} f\left(\frac{W}{L}\right)^6 + \frac{7 V_R V_D^{\,6}}{12} f\left(\frac{W}{L}\right)^7 + \frac{V_D^{\,7}}{14} f\left(\frac{W}{L}\right)^8 = \frac{W^6}{1.8 \times 10^{-35}}$$

(2.20)

where all voltages are in V while L and W are in cm. For design of optimum pillar parameters, the condition $E_{min} = 0$ is imposed to ensure complete depletion at $V_R = V_{BR}$. Then, (2.19) yields

$$V_D = \frac{V_{BR}}{f(W/L)}$$

(2.21)

Substitute (2.21) in (2.20) to yield

$$1 + \frac{18239}{840} f\left(\frac{W}{L}\right) = \frac{L^6}{1.8 \times 10^{-35} V_{BR}^{\,7}}$$

(2.22)

21

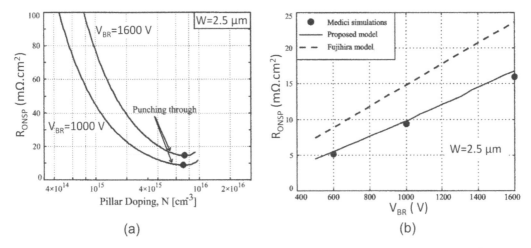

Fig. 2.5 (a) The simulated R_{ONSP} as a function of N for V_{BR} = 1000, 1600 V and W = 2.5 μm. Dark circles indicate the minimum simulated R_{ONSP} while arrows indicate that predicted by the model [19]. (b) Comparison of the R_{ONSP} versus V_{BR} predicted by [19] and [17] for a Si superjunction (reproduced from [19]).

Equations (2.18), (2.21) and (2.22) are numerically solved to obtain the L, N and W parameters for a balanced silicon superjunction. Figure 2.5(a) plots the simulated R_{ONSP} as a function of N for a fixed V_{BR}. The pillar doping corresponding to the minimum R_{ONSP} is the optimum point. In contrast, the model predicts a doping, at least 30 % lesser than the optimum point as show in Fig. 2.5(a). Nevertheless, it predicts an improved V_{BR}-R_{ONSP} trade-off compared to the prior model reported in Ref. [17] as shown in Fig. 2.5(b).

Limitations: The work involves rigorous mathematical derivations and yet doesn't yield the minimum R_{ONSP} for a target V_{BR} predicted by TCAD simulations. The method do not give closed form equations; instead, three equations are to be numerically solved to obtain the pillar parameters. The equations give a distorted insight that N_{opt} depends on L_{opt} and W_{opt} whereas it is a function of W_{opt} alone. Further, the R_{ONSP} calculations are made by neglecting the doping dependency on mobility and zero bias depletion width

2.2.4 Balanced Yu Method

This is the first work that reports the design of a 4H-SiC superjunction device [20]. Unlike the prior works, the breakdown electric field profile along the critical path is modelled as a polynomial fit of the form

$$E = \frac{1}{2}\left(E_1 + \mid E_1 \mid\right)$$

(2.23)

$$E_1 = \begin{cases} E_f + \dfrac{qN}{\varepsilon}\dfrac{(y_f - y)^m}{m.y_f^{m-1}} & y \in \left[0, y_f\right] \\[4mm] E_f & y \in \left[y_f, L - y_f\right] \\[4mm] E_f - \dfrac{qN}{\varepsilon}\dfrac{(y - (L - y_f))^m}{m.y_f^{m-1}} & y \in \left[L - y_f, L\right] \end{cases}$$

(2.24)

where

$$E_f = V_{BR}/L \qquad y_f = 2.4W \qquad m = 3.2$$

(2.25)

This is compared with numerical simulation in Fig. 2.6(a). These expressions (2.23)- (2.25) are to be used for the evaluation of the ionization integral,

$$\int 4 \times 10^{-48} E^8 dy = 1$$

(2.26)

Due to the complex nature of the electric field function, it is impossible to obtain a closed form expression for V_{BR} from the ionization integral. Hence, this task has to be done numerically by

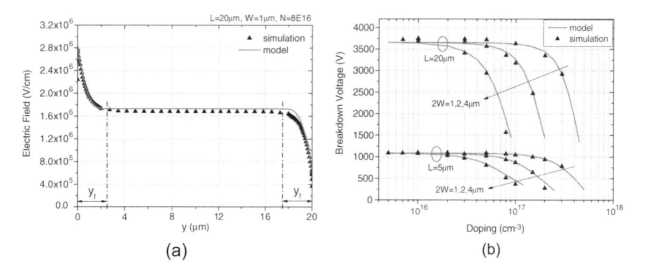

Fig. 2.6 (a) Breakdown electric field distribution model compared to simulation data: $L = 20\ \mu m$, $W = 1\ \mu m$, and $N = 8 \times 10^{16}\ cm^{-3}$. (b) The V_{BR} model compared to numerical simulation results for various L, W, and doping conditions for 4H-SiC superjunction devices (reproduced from [20]).

assuming various values of V_{BR} until Eq. (2.26) is satisfied. The results of this model for various values of pillar parameters are compared with numerical simulations and plotted in Fig. 2.6(b). The R_{ONSP} of the device is calculated using (2.42) and (2.43). The models (2.23)-(2.26) and (2.42)-(2.43) are used to arrive at the optimum pillar parameters that yield a target V_{BR} and minimum R_{ONSP}. Due to the absence of any direct analytical equation for the design, the parameters, L, N and W have to be varied across a wide range of values by maintaining a fixed V_{BR} and ultimately choose the combination that yields the minimum R_{ONSP}.

Limitations: This method involves numerical evaluation of the ionization integral and as a result do not yield any analytical equations for the design of pillar parameters. The model may be used for determining the V_{BR} and R_{ONSP} for a general L, N and W. But, the reverse problem of designing the pillar parameters for a target V_{BR} and minimum R_{ONSP} is possible only by extensive computation involving varying L, N and W over a wide range of values.

2.2.5 Optimum p- Pillar Aspect Ratio

Ref. [17] derived a relation between R_{ONSP} and W given by (2.10); One may infer from this equation that R_{ONSP} of an SJ may be reduced indefinitely by reducing W or equivalently increasing the pillar aspect ratio, $r = L/2W$ for a given V_{BR}. In fact, prior works [17]-[20] have suggested to choose the highest r realizable by technology for achieving the minimum R_{ONSP}. Literature [21],[22] has reported pillar aspect ratios up to 25 (10) in Si (4H-SiC) SJs.

However, (2.10) has two shortcomings that make the above-mentioned inference to be erroneous when applied for practical devices. Firstly, it is applicable only to an ideal balanced SJ; whereas, real SJ is prone to inevitable charge imbalance due to process variations and has to be considered while deriving any realistic design guideline. Secondly, it assumed that the entire n-pillar width is available for conduction; whereas in practice, the effective width available for conduction is only $W- W_d$, where W_d is the zero-bias n-pillar depletion width of the junction between p and n pillars.

Recently, Ref. [23] studied a balanced SJ to solve the second shortcoming mentioned above. It showed that due to the presence of W_d, there exist a material dependent minimum value of W, W_0 below which R_{ONSP} can no longer be reduced. It estimated a U-shaped variation of R_{ONSP} with W of a balanced SJ, from which the minimum R_{ONSP} can be located at $W_0 = 0.09$

(0.025) µm in Si (4H-SiC) SJs. Using these and typical $L \approx 4$–61 (5–75) µm for $V_{BR} = 0.1$–1 (1–10) kV it can be shown that the optimum r, $r_0 = 22$–340 (100–1500). Ref. [24] gave a MATLAB based optimization method for balanced Si SJ design as per which R_{ONSP} was found to fall up to $r = 70$ (calculations for higher r were not provided).

However, no study of R_{ONSP} with W or r has been conducted so far that accounts the effect of charge imbalance due to process variations.

2.3 IMBALANCED SUPERJUNCTION DESIGN

The major source of charge imbalance is due to the mismatch between the p- and n-pillar doping primarily due to the partial activation of p-dopants during annealing; the effect being more dominant in 4H-SiC than Si. The secondary source of charge imbalance is due to the difference is pillar widths due to lithographic constraints. For analysis purpose, literature assumes $W_n = W_p$ and $N_d > N_a$ (see Fig. 1.4(a)). The presence of charge imbalance introduces a slope in the breakdown electric field distribution along the critical path which degrades the V_{BR} as shown in Fig. 1.4(b). Hence, design equations that can precisely predict the V_{BR} and R_{ONSP} of a superjunction device in the presence of charge imbalance and that can yield the optimum pillar parameters for a target V_{BR} are required. Some attempts to this are summarized below.

2.3.1 Napoli Method

This is the first work that reports the modelling of electric field and breakdown voltage of imbalanced superjunction devices [25]. The proposed model uses a superposition approach that treats the imbalanced superjunction device as the superposition of a balanced superjunction with $N = (N_d + N_a)/2$ and a PiN diode with $N^* = (N_d - N_a)/2$ n-type doping. Due to this superposition treatment, the electric field along the critical path ($x = \pm W$) of an imbalanced superjunction, $E_{SJ}(y)$ is the sum of the balanced component, $E_{bal}(y)$ and the differential component, $E_{diff}(y)$.

$$E_{SJ}(y) = E_{bal}(y) + E_{diff}(y) \tag{2.27}$$

Putting $E_{bal}(y)$, obtained by solving the 2-D Poisson equation for a balanced SJ, and a triangular field profile for $E_{diff}(y)$ in (2.27) yields,

$$E_{SJ}(y) = \frac{2V_u}{L}\left(1-\frac{y}{L}\right) + \frac{V_B}{L} + \frac{V_D}{L}\left(1-\frac{2y}{L}\right) + \sum_{n=1}^{\infty} \frac{K_n \gamma_n}{2} \frac{\cos(K_n y)}{\cosh(K_n W)} \tag{2.28}$$

where

$$V_u = \frac{qN^*}{2\varepsilon}L^2 \qquad V_D = \frac{qN_d}{2\varepsilon}L^2 \tag{2.29}$$

Using an E_C based approach, assume that V_{BR} corresponds to the condition for which $E_{SJ}(y) = E_C$. This model is compared with numerical simulation for various values of imbalance level, CI = $(1-N_a/N_d)\times100$ %, in Fig. 2.7(a) (reproduced from [25]). Integration of (2.28) along the critical path yields $V_R = V_B + V_u$ where V_R is the applied reverse bias and V_D, V_u are the voltage drop due to balanced and differential component respectively. Further, assume that the device undergoes punch-through prior to the onset of avalanche breakdown. Simplification of the above equations yields the V_{BR} of a punch-through superjunction with charge imbalance and is given by

$$V_{BR} = LE_C - V_u - V_D f(W/L) \tag{2.30}$$

The V_{BR} as a function of charge imbalance level for two different n-pillar doping values given by the model (2.30) and that obtained using numerical simulations is given in Fig. 2.7(b) (reproduced from [25]).

Limitations: Although E_C depends on both doping and geometry, this method suggests assuming a typical value of E_C, which can cause significant error in predicting the V_{BR}. The

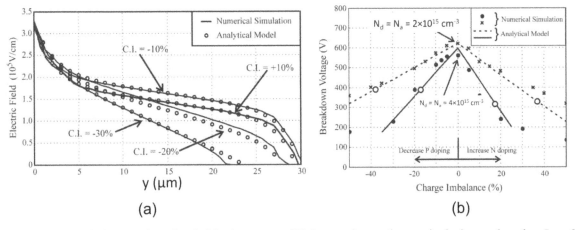

(a)   (b)

Fig. 2.7 Breakdown electric field along $x = W$ for various charge imbalance levels, $L = 30$ μm and $W = 2.5$ μm $N_d = 4 \times 10^{15}$ cm^{-3}. (b) The V_{BR} for imbalanced superjunction structure as a function of doping imbalance between the p- and n- pillars (reproduced from [25]).

model presented in this work is unsuitable for estimating the V_{BR} of a superjunction which has undepleted pillar at breakdown, i.e. a non-punch-through device. It has complex mathematical functions involved in the calculations and hence, the reverse problem of estimating the pillar parameters that yield a target V_{BR} is not straight forward.

2.3.2 Improved Napoli Method

The prior model [25] was improvised in this work to account for non-punch through devices. A double exponential electric field model was suggested as prior models showed significant deviation from simulated values towards the bottom of the critical path, near $y = L$.

$$E_{SJ}(y) = \frac{2V_u}{L}\left(1 - \frac{y}{L}\right) + \frac{V_B}{L} + \frac{V_{SJ}}{L}\exp\left(-\frac{2y}{f(W/L)L}\right) - \frac{V_{SJ}}{L}\exp\left(\frac{2(y-L)}{f(W/L)L}\right) \tag{2.31}$$

where $V_{SJ} = V_D \times f(W/L)$, V_D and V_u are from (2.29).

For the punch through device, put $E_{SJ}(0) = E_C$ and integrate along the critical path from $y = 0$ to L. This yields punch through breakdown voltage, $V_{BR,PT}$

$$V_{BR,PT} = LE_C - V_u - V_D f(W/L) \tag{2.32}$$

To identify the boundary of punch through and non punch through, we need to use the conditions, $E_{SJ}(0) = E_C$ and $E_{SJ}(L) = 0$. This yields the condition,

$$V_u + V_{SJ} - V_{SJ}\exp\left(-\frac{2}{f(W/L)}\right) = \frac{E_C L}{2} \tag{2.33}$$

and the corresponding breakdown voltage,

$$V_{BR,NPT-PT} = \frac{LE_C}{2} - V_u \tag{2.34}$$

For non punch through devices it is crucial to identify y^* along the critical path such that $E_{SJ}(y^*) = 0$. This yields the condition,

$$y^* = L - L\frac{f^2(V_u + V_{SJ}/f)}{V_{SJ}}\left(\beta + \beta^2\right) \tag{2.35}$$

27

where

$$\beta = NP\frac{V_{SJ}(V_{SJ}+V_u)}{f^2(V_u+V_{SJ}/f)^2} \qquad NP = 1-\frac{E_C L/2}{V_{SJ}+V_u} \tag{2.36}$$

For non-punch through device, put $E_{SJ}(0) = E_C$, $E_{SJ}(y^*) = 0$ and integrate along the critical path from $y = 0$ to y^*. This yields non-punch through breakdown voltage, $V_{BR,NPT}$

$$V_{BR,NPT} = V_{BR,PT} - \left[V_{BR,PT} - \frac{V_u y^*}{L}\right]\left(1-\frac{y^*}{L}\right) \tag{2.37}$$

Limitations: The only improvement in this work over the Napoli method is that it can now predict the V_{BR} of a punch through and non-punch through device. All other drawbacks discussed for the Napoli method is applicable here too. The V_{BR} as a function of charge imbalance level, given by this model (2.37), the prior model (2.30) and that obtained using numerical simulations is given in Fig. 2.8 (reproduced from [26]) for comparison purpose.

2.3.3 Imbalanced Yu Method

This work reports the modelling of 4H-SiC superjunction devices with charge imbalance. The

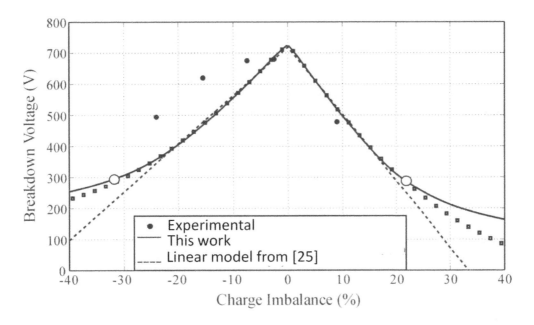

Fig. 2.8 The V_{BR} as a function of doping imbalance between the p- and n- pillars for an imbalanced superjunction structure with $L = 40 \ \mu$m and $W = 4 \ \mu$m (reproduced from [26]).

breakdown electric field along the critical path is given as

$$E = \frac{1}{2}\left(E_1 + |E_1|\right) \qquad (2.38)$$

$$E_1 = \begin{cases} E_{slope} + \dfrac{q*\max(N_d, N_a)}{\varepsilon} \dfrac{(y_f - y)^m}{m.y_f^{m-1}} & y \in \left[0, y_f\right] \\ E_{slope} & y \in \left[y_f, L - y_f\right] \\ E_{slope} - \dfrac{q*\max(N_d, N_a)}{\varepsilon} \dfrac{(y - (L - y_f))^m}{m.y_f^{m-1}} & y \in \left[L - y_f, L\right] \end{cases} \qquad (2.39)$$

where

$$E_{slope} = \frac{V_R}{L} - \frac{q|N_d - N_a|}{2\varepsilon}\left(y - \frac{L}{2}\right) \qquad (2.40)$$

and

$$E_f = V_{BR}/L \qquad y_f = 2.4W \qquad m = 3.2 \qquad (2.41)$$

This model is compared with the results obtained using numerical simulations and is given in Fig. 2.9(a) (reproduced from [20]). These derived expressions, (2.38)-(2.41) are to be used for

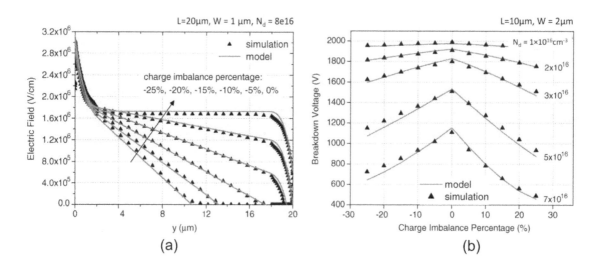

Fig. 2.9 (a) The breakdown electric field distributions in charge-imbalanced superjunction diode structures with various charge imbalance percentages; comparison of the proposed model with simulations. (b) Charge imbalance model compared to simulations: The breakdown voltage versus charge imbalance percentage at various doping concentrations (reproduced from [20]).

the evaluation of the ionization integral. This model would not result in any analytical equation and hence numerical methods are to be used. The V_{BR} so predicted using this method as a function of charge imbalance level is plotted in Fig. 2.9(b) (reproduced from [20]).

The R_{ONSP} of the device is calculated by accounting for doping dependency of mobility and zero bias n-pillar depletion width and is given by

$$R_{ONSP} = \frac{L}{qN_d\mu_n(N_d)} \times \frac{2W}{W - W_d} \tag{2.42}$$

where

$$\mu_n(N_d) = 40 + \frac{950 - 40}{1 + \left(\dfrac{N_d}{2 \times 10^{17}}\right)^{0.76}} \ \text{cm}^2/\text{Vs}, \ W_d = \sqrt{\frac{\varepsilon V_{bi}}{qN_d}} \tag{2.43}$$

where N_d is in cm^{-3}. A trial and error approach is suggested to design L, N_d and W_n of a superjunction for a target V_{BR} and minimum R_{ONSP} for a specified level of charge imbalance. Vary L, N_d and imbalance level across a range of typical values for a given W_n and V_{BR} and plot R_{ONSP} and N_d as a function of imbalance level with L as a parameter. Choose the combination of L, N_d that yields the minimum R_{ONSP} for the imbalance level expected.

Limitations: This method involves numerical evaluation of the ionization integral and as a result do not yield any analytical equations for the design of pillar parameters. The model may be used for determining the V_{BR} and R_{ONSP} for a general L, N_d and W_n. But, the reverse problem of designing the pillar parameters for a target V_{BR} and minimum R_{ONSP} is possible only by trial and error. The charge imbalance due to error in trench width during lithography and etching is not considered. Instead, only the contribution of mismatch in p- and n-pillar doping is considered.

2.3.4 Alam Method

This work reports a lengthy yet systematic method to arrive at the optimum pillar parameters of an imbalanced superjunction [27], [28]. The charge imbalance, CI = (1-N_d/N_a)×100 %, provided $N_d > N_a$. The condition adopted for arriving at the optimum pillar parameters is the maximization of the figure of merit (FOM), V_{BR}^2/R_{ONSP}. For determining the

V_{BR} for a given L, N_a, N_d and W_n, the ionization integral,

$$\int a_p \exp\left(-\int (a_p - a_n)dy\right)dy = 1 \tag{2.44}$$

has to be evaluated where,

$$a_{n,p} = k_{1\_n,p}\left(E - E_{m\_n,p}\right)k_{2\_n,p} \tag{2.45}$$

and k_1, k_2 and E_m are fitting parameters [29]. Here, electric field along critical path, E is given by

$$E(y) = \pm\frac{V_u}{L}\left(1-2\frac{y}{L}\right) + \frac{V_R}{L} \pm \frac{V_{D,A}}{L}\left(1-\frac{2y}{L}\right) \pm 4\left(\frac{V_A + V_D}{L}\right) \times \sum_{n=1}^{\infty} \frac{\left((-1)^n - 1\right)\cos(n\pi y/L)\sinh(n\pi W/L)}{(n\pi)^2 \sinh(2n\pi W/L)} \tag{2.46}$$

where

$$V_u = \left(qN/2\varepsilon\right)L^2; \quad V_{D,A} = \left(qN_{D,A}/2\varepsilon\right)L^2; \quad N = \left(N_d - N_a\right)/2; \quad N_{D,A} = \left(N_d + N_a\right)/2 \tag{2.47}$$

Equations (2.44)-(2.47) have to be solved numerically with varying V_R until the integral in (2.44) is evaluated to be unity. The V_R value that satisfies (2.46) is the V_{BR} of the device. Further, it reports an eight-step iterative procedure to find the drift layer thickness that gives the optimum FOM for a given level of imbalance, doping, and pillar width. Following this method,

$$L_{opt} = \frac{2\varepsilon E_{C,eff}}{3qN} \tag{2.48}$$

where $E_{C,eff}$ is the extrapolated peak of the linear portion of the breakdown electric field distribution as shown in Fig.2.5(b) and whose value is yielded by the eight step procedure. A systematic procedure to design an imbalanced superjunction is discussed below. Although the procedure works for all semiconductors, calculations are presented for 4H-SiC material.

Design Procedure:

1) A typical value of W is assumed and N is varied over an arbitrary range of values. For each N, optimum L is calculated using (2.48) following the eight step procedure. The Fig. 2.10(a) shows the R_{ONSP} versus V_{BR} for various values of pillar doping or equivalently dose. The optimum FOM points for each dose is marked.

2) Plot the optimum FOM values from step (1) as a function of dose to see the dose

31

corresponding to the highest value of FOM (see Fig. 2.10(b)).

3) Repeat steps (1) and (2) for various values of W across an arbitrary range and plot FOM versus V_{BR} (see Fig. 2.10(c)). Choose the L, N and W corresponding to the highest FOM for the target V_{BR}.

The R_{ONSP} versus V_{BR} of the devices designed using the above method for various values of charge imbalance level is plotted in Fig. 2.10(d).

Limitations: The suggested procedure is computationally intensive as it suggests to vary W, N and L across a wide range of values and ultimately choose the combination that maximize the function, V_{BR}^2/R_{ONSP}. For each of these values, the ionization integral needs to be evaluated several times by assuming various values of V_R until the integral value is unity, making it further challenging. Adding to this is the lack of guidelines for choosing initial/final values for each of

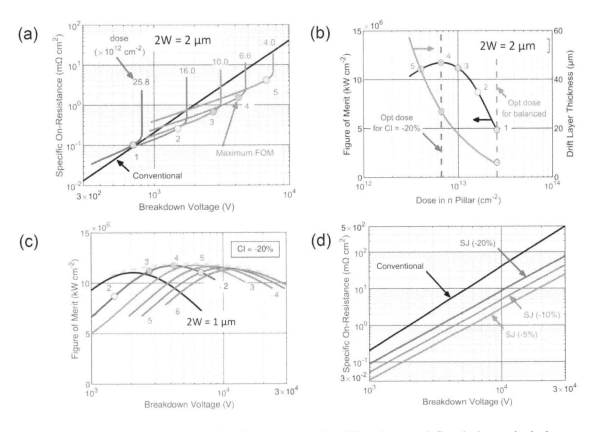

Fig. 2.10 (a) The R_{ONSP} versus V_{BR} for a range of n-pillar dose and fixed charge imbalance and pillar width (b) FOM versus n-pillar dose values obtained from (a). (c) FOM versus V_{BR} for various values of W. (d) The R_{ONSP} versus V_{BR} for various charge imbalance levels for a 4H-SiC superjunction. (reproduced from [28])

the iterating parameters. A poor choice of initial guess can add significant amount of computation time. Further, this method ultimately demands fabrication of trench with unrealistic aspect ratio for low imbalance levels, higher than that reported for 4H-SiC. It superficially deals with charge imbalance arising when n-pillar dose is lower than p-pillar and neglects the effect of charge imbalance due to error in trench width during lithography and etching while explaining the design procedure.

2.4 THERMAL CONSIDERATIONS IN DRIFT LAYER DESIGN

It is desirable for all power semiconductor devices to be able to reliably operate at elevated temperatures. In fact, the promise of SiC in power applications boils down largely to its ability to handle higher temperature operation. However, all reports in the literature [17]–[20], [23]–[28] that discuss novel drift layer design strategies of power devices assume room temperature for all calculations. This is because, for all practical purposes, drift layer design do not decide the maximum operating temperature, T_{max}, of power devices. In fact, Ref. [30] discusses the various factors that limit T_{max} of Si and SiC unipolar devices and also suggests a few techniques to increase the same; none of these techniques involve modification of the drift layer design. Based on the above, all the calculations reported in this thesis will be at room temperature.

2.5 OBJECTIVES OF THE THESIS

The above review has established that the solutions available for the design of both ideal balanced and as well as practical imbalanced SJ has significant scope for improvement. The state of the art closed form solution for a balanced SJ is sub-optimal as it predicts ~ 30 % lower R_{ONSP} for a given V_{BR} than that predicted by TCAD simulations. On the other hand, no closed form solutions are available so far for the design of a practical imbalanced SJ. Also, the optimum pillar aspect ratio that yields the minimum R_{ONSP} has not been studied so far in the presence of charge imbalance. Further, the fabrication of SJ in 4H-SiC material has been problematic due to the difficulties associated with fabricating the p- pillar as explained in Chapter 1.

In the above perspective, the objectives of our work are as follows.

• Derive material independent closed-form solutions for the optimum pillar parameters of a balanced SJ that yield the minimum R_{ONSP} for a target V_{BR} and that match with TCAD simulations.

- Derive material independent closed-form solutions for the optimum pillar parameters of an imbalanced SJ that yield the minimum R_{ONSP} for a target V_{BR} and that match with TCAD simulations. Also, study the R_{ONSP} versus r behavior for the first time and derive a simple equation for the optimum r as a function of target V_{BR} and material.

- Study the practicability of Charge Sheet Superjunction (CSSJ) in 4H-SiC material to solve the p- pillar fabrication difficulty of SJ; CSSJ being a novel device concept proposed by our group a decade ago in the context of Si material. Derive material independent analytical solution for the design of CSSJ for a target V_{BR} in the range $0.1 - 10$ kV and show that apart from the fabrication related advantages, CSSJ also offers a lower R_{ONSP} than SJ.

2.5 REFERENCES

[1] H. Wang, C. Wang, B. Wang, H. Long and K. Sheng, "Hybrid termination with wide trench for 4H-SiC super-junction devices", *IEEE Electron Device Lett.*, vol. 42, no. 2, pp. 216-219, Feb. 2021.

[2] T. Masuda, T. Hatayama, S. Harada and Y. Saitou, "Edge termination design with strong process robustness for 1.2 kV-class 4H-SiC super junction V-groove MOSFETs", *Proc. 32nd Int. Symp. Power Semiconductor Devices ICs*, pp. 166-169, Sep. 2020.

[3] C. H. Cheng, C. F. Huang, K. Y. Lee and F. Zhao, "A novel deep junction edge termination for superjunction MOSFETs", *IEEE Electron Device Lett.*, vol. 39, no. 4, pp. 544-547, Apr. 2018.

[4] F. Udrea, T. Trajkovic, J. Thomson, L. Coulbeck, P.R. Waind, G.A.J. Amaratunga and P. Taylor, "Ultra-high voltage device termination using the 3D RESURF (super-junction) concept—Experimental demonstration at 6.5 kV", *Proc. 13th Int. Symp. Power Semiconductor Device ICs*, pp. 129-132, Jun. 2001.

[5] B. J. Baliga, Fundamentals of power semiconductor devices, Springer Publishers, 2008.

[6] C. Hu, "A parametric study of power MOSFET's", *IEEE Power Electronics Specialists Conf.*, pp. 385-395, 1979.

[7] Z. Li, V. Pala, and T. P. Chow, ``Avalanche breakdown design parameters in GaN," *Jpn. J. Appl. Phys.*, vol. 52, no. 8, May 2013, doi: 10.7567/JJAP.52.08JN05.

[8] Y. Zhang, M. Sun, D. Piedra, J. Hu, Z. Liu, Y. Liu, X. Gao, K. Shepard, and T. Palacios , "1200 V GaN vertical fin power field-effect transistors", *IEDM Tech. Dig.*, pp. 9.2.1-9.2.4, Dec. 2017, doi: 10.1109/IEDM.2017.8268357.

[9] G. Brezeanu, A. Visoreanu, M. Brezeanu, F. Udrea, G.A.J. Amaratunga, I. Enache, I. Rusu and F. Draghici, "Off - State Performances of Ideal Schottky Barrier Diodes (SBD) on Diamond and Silicon Carbide", *International Semiconductor Conference*, vol. 2, pp. 319-322, Sept. 2006.

[10] I. C. Kizilyalli, A. P. Edwards, O. Aktas, T. Prunty, and D. Bour, ``Vertical power p-n diodes based

on bulk GaN," *IEEE Trans. Electron Devices*, vol. 62, no. 2, pp. 414_422, Feb. 2015. doi: 10.1109/TED.2014.2360861.

[11] N. Allen, M. Xiao, X. Yan, K. Sasaki, M. J. Tadjer, J. Ma, R. Zhang, H. Wang, Y. Zhang, "Vertical Ga_2O_3 Schottky barrier diodes with small-angle beveled field plates: A Baliga's figure-of-merit of 0.6 GW/cm^2 ",*IEEE Electron Device Lett.*, vol. 40, no. 9, pp. 1399-1402, Sep. 2019, doi: 10.1109/LED.2019.2931697.

[12] R. Singh, J. A. Cooper, M. R. Melloch, T. P. Chow and J. W. Palmour, "SiC power Schottky and PiN diodes",*IEEE Trans. Electron Devices*, vol. 49, no. 4, pp. 665-672, Apr. 2002, doi: 10.1109/16.992877.

[13] H. Huang, J. Huang, H. Hu, J. Cheng and B. Yi, "Analytical models of breakdown voltage and specific on-resistance for vertical GaN unipolar devices", *IEEE Access*, vol. 7, pp. 140383-140390, 2019, doi: 10.1109/ACCESS.2019.2944028.

[14] G. Chicot, D. Eon and N. Rouger, "Optimal drift region for diamond power devices", *Diamond Rel. Mater.*, vol. 69, pp. 68-73, Oct. 2016, doi: https://doi.org/10.1016/j.diamond.2016.07.006.

[15] C. Hu, "Optimum doping profile for minimum ohmic resistance and high breakdown voltage", *IEEE Trans. Electron Devices*, vol. ED-26, pp. 243, 1979.

[16] D. P. Bertsekas, Constrained Optimization and Lagrange Multiplier Methods, Academic Press, 1982.

[17] T. Fujihira, "Theory of semiconductor superjunction devices," *Jpn. J. Appl. Phys.,* vol. 36, no. 10, pp. 6254–6262, Oct. 1997.

[18] X. B. Chen, P. A. Mawby, K. Board, and C. A. T. Salama, "Theory of a novel voltage-sustaining layer for power devices," *Microelectron. J.*, vol. 29, no. 12, pp. 1005–1011, Dec. 1998.

[19] G. M. Strollo and E. Napoli, "Optimal on-resistance versus breakdown voltage tradeoff in superjunction power devices: A novel analytical model," *IEEE Trans. Electron Devices*, vol. 48, no. 9, pp. 2161–2167, Sept. 2001.

[20] L. Yu and K. Sheng. "Modeling and optimal device design for 4H-SiC super-junction devices," *IEEE Trans. Electron Device,* vol. 55, no. 8, pp. 1961-1969, Jul. 2008.

[21] J. Sakakibara, Y. Noda, T. Shibata, S. Nogami, T. Yamaoka and H. Yamaguchi, "600 V-class super junction MOSFET with high aspect ratio P/N columns structure", *Proc. 20th ISPSD*, pp. 299-302, May 2008, doi: 10.1109/ISPSD.2008.4538958.

[22] R. Kosugi, S. Ji, K. Mochizuki, K. Adachi, S. Segawa, Y. Kawada, et al., "Breaking the theoretical limit of 6.5 kV-class 4H-SiC super-junction (SJ) MOSFETs by trench-filling epitaxial growth", *Proc. 31st Int. Symp. Power Semicond. Devices ICs (ISPSD)*, pp. 39-42, May 2019. doi: 10.1109/ISPSD.2019.8757632.

[23] H. Kang and F. Udrea, "True material limit of power devices—Applied to 2-D superjunction MOSFET," *IEEE Trans. Electron Devices*, vol. 65, no. 4, pp. 1432–1439, Apr. 2018, doi: 10.1109/TED.2018.2808181.

[24] H. Huang, S. Xu, W. Xu, K. Hu, J. Cheng, H. Hu, and B. Yi, "Optimization and Comparison of Drift Region Specific ON-Resistance for Vertical Power Hk MOSFETs and SJ MOSFETs With Identical Aspect Ratio", *IEEE Trans. Electron Devices*, vol. 67, no. 6, pp. 2463–2470, Jun. 2020, doi: 10.1109/TED.2020.2989418.

[25] E. Napoli, H.Wang, and F. Udrea, "The effect of charge imbalance on superjunction power devices: An exact analytical solution," *IEEE Electron Device Lett.*, vol. 29, no. 3, pp. 249–251, Mar. 2008.

[26] H. Wang, E. Napoli, and F. Udrea, "Breakdown voltage for superjunction power devices with charge imbalance: An analytical model valid for both punch through and non punch through devices," *IEEE Trans. Electron Devices*, vol. 56, no. 12, pp. 3175–3183, Dec. 2009.

[27] M. Alam, D. T. Morisette, and J. A. Cooper, "Practical design of 4H-SiC superjunction devices in the presence of charge imbalance," in *Proc. Int. Conf. Silicon Carbide Related Mater.*, 2017.

[28] M. Alam, D. T. Morisette, and J. A. Cooper, "Design guidelines for superjunction devices in the presence of charge imbalance." *IEEE Trans. Electron Devices,* vol. 65, no. 8, pp. 3345-3351, Aug. 2018.

[29] T. Hatakeyama, T. Watanabe, T. Shinohe, K. Kojima, K. Arai, and H. Sano, "Impact ionization coefficients of 4H silicon carbide," *Appl. Phys. Lett.,* vol. 85, no. 8, pp. 1380-1382, 2004.

[30] K. Sheng, "Maximum junction temperatures of SiC power devices", *IEEE Trans. Electron Devices,* vol. 56, no. 2, pp. 337-342, Feb. 2009.

CHAPTER 3

BALANCED SUPERJUNCTION DESIGN

A balanced superjunction (SJ) is an ideal device which has equal amount of charge in n- and p- pillars; this demands zero error in the fabrication processes and thus not practical. However, theoretical studies often consider a balanced SJ for modeling and design primarily because of two reasons. Firstly, analysis of balanced SJ yields the theoretical lower limit of R_{ONSP} for a given V_{BR}; comparing this with the measured R_{ONSP} and V_{BR} serves as a guideline for the experimentalists to assess the scope for improving their process. Secondly, analysis of balanced SJ is easier and more intuitive as compared to a practical SJ where charge imbalance between the n- and p- pillars has to be considered. Hence, this chapter considers the modeling and design of an ideal balanced SJ while the next chapter considers design of a practical imbalanced SJ. This chapter is based on the work published in Ref. [1] and a portion of Ref. [2].

Analytical derivation of the optimum pillar parameter values L_{opt}, W_{opt} and N_{opt} of a balanced SJ (see Fig. 1.4(a)) that yields the lower limit of R_{ONSP} achievable for a target V_{BR} has been a long standing problem. There is significant scope for improving the prior solutions [3], [4]-[6]. Ref. [3], [4] yield V_{BR} and R_{ONSP} for a general L, W and N; the process of using them to get L_{opt}, W_{opt} and N_{opt} is physically opaque and computationally intensive. Ref. [5], [6] improve upon [3] and directly give an analytical solution for L_{opt}, W_{opt} and N_{opt} (see Appendix). However, as acknowledged in [5], these solutions are "suboptimal," since their predicted N_{opt} is lower and hence the minimum R_{ONSP} higher than those obtained from accurate numerical calculations by > 30%. Further, as we show, the impression given by the solution that L_{opt} depends on W_{opt}, and N_{opt} on L_{opt}, is distorted. Moreover, the solution needs iterative calculations. Nevertheless, being the state of the art guideline, originally proposed for silicon material [5], it is used for even SiC [7] and GaN [8] SJ modeling after making material specific changes.

In section 3.1, we derive simple closed-form solutions for L_{opt}, W_{opt} and N_{opt}. Our estimates of N_{opt} are higher and hence the R_{ONSP} lower than in prior works by > 30 %. Moreover, our solutions have a generic form that permits quick calculations across different materials. They are validated by TCAD simulation [9] for low and wide bandgap materials – Si and 4H-SiC. Their

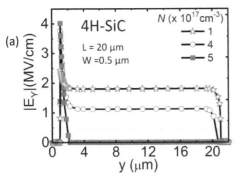
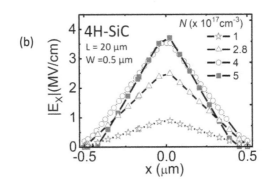

Fig. 3.1 Simulated field distribution at breakdown in a 4H-SiC superjunction. (a) Vertical field over pillar length at $x = W$. (b) Lateral field over pillar width at $y = L/2$.

results for GaN and diamond are given to reveal the wide 100 V – 30 kV application range of devices in these materials. In Section 3.2, our design procedure is illustrated by showing the calculations for a 4H-SiC SJ with $V_{BR} = 3$ kV. In Section 3.3, our solutions are used to reveal the scope for reducing the R_{ONSP} of several fabricated devices reported in literature by process improvement.

3.1 OPTIMUM PILLAR PARAMETERS

The difficulties of Ref.[4]-[6] in deriving the optimum parameters arise from their approach of developing the design equations starting from a model of the 2-D field distribution for a general N. Unfortunately, for moderate to high N, the distribution of E_y over L along $x = \pm W$ contains rapidly varying segments near the p^+n and nn^+ junctions. This is shown in Fig. 3.1(a) for a SiC super-junction with $N \geq 4 \times 10^{17}$ cm^{-3}, but holds true for Si, GaN and diamond super-junctions as well, as shown by our calculations. Such segments can only be modeled using infinite series or power law function of N, W and L which are difficult to carry through the ionization integral analytically.

However, we recognize that an analysis of the field distribution near N_{opt} alone suffices for this purpose. With this realization, we adopt a different approach and achieve simple closed-form solutions for L_{opt}, W_{opt} and N_{opt} as explained below.

3.1.1 Optimum Pillar Doping

Refer to Fig. 3.2(a) showing the doping dependence of various voltages associated with the SJ structure. Let V_{SJ} denote the value of V_{BR} for low N, V_{p+n} the breakdown voltage of the 1-D p^+n

junction, and

$$V_{pn} = qNW^2/\varepsilon_s \tag{3.1}$$

the voltage at which pillars are fully depleted by lateral triangular field distribution. At low N, the pillars get depleted fully *well before* the applied voltage reaches V_{p+n}, i.e. we have $V_{pn} \ll V_{p+n}$. Under this condition, for applied voltage $> V_{pn}$, most vertical field lines emanating from the n$^+$ region terminate on the p$^+$ region, resulting in an approximately uniform vertical field over the pillar length L. The $V_{BR} = V_{SJ} \gg V_{pn}$ is the area under this uniform field distribution when the field attains a critical value. We have $V_{SJ} \gg V_{p+n}$ since V_{p+n} is the area under a field distribution which decreases linearly from the critical value to zero over a depletion width $\ll L$. The SJ operation and hence the $V_{BR} = V_{SJ}$ condition is maintained up to an N for which the pillars get depleted fully *just before* the applied voltage reaches V_{p+n}, i.e. for which

$$V_{pn} = V_{p+n}. \tag{3.2}$$

Hence, the minimum R_{ONSP} is obtained for the optimum doping N_{opt} which satisfies (3.2). For $N > N_{opt}$, at applied voltage $= V_{p+n}$, the pillars are only partially depleted, e.g. see curves for $N = 4$, 5×10^{17} cm^{-3} in Fig. 3.1(b). Hence, the device breaks down at the pillar edges near the n$^+$p and p$^+$n junctions. Thus, the device V_{BR} degrades from V_{SJ} to V_{p+n}.

Unlike in the above simplistic explanation, the actual transition of the V_{BR} from V_{SJ} to V_{p+n} is not abrupt but smooth (see Fig. 3.2(a)), so that at $N = N_{opt}$, $V_{p+n} < V_{BR} < V_{SJ}$ for the following reasons: $V_{BR} < V_{SJ}$ since non-uniformity sets into the vertical field distribution over pillar length L just below N_{opt}, because some of the end to end field lines terminate into or emanate from the pillars; $V_{BR} > V_{p+n}$ since just above N_{opt}, the pillars partially depleted near the middle remain fully depleted near the n$^+$ and p$^+$ ends where the field is 2-dimensional. We can derive N_{opt} of an SJ in any material from (3.1) and (3.2) if we have a V_{p+n} formula that works across materials. We develop the required V_{p+n} formula by generalizing the approach used in literature for specific materials. Assume impact ionization dominated breakdown, which occurs when the ionization integral over the depletion width y_d along the critical path $x = \pm W$ approaches unity [10],[11], i.e

$$I = \int_0^{y_d} \alpha_{eff} dy \rightarrow 1 \tag{3.3}$$

where α_{eff} is the effective impact ionization coefficient of electrons and holes. To facilitate

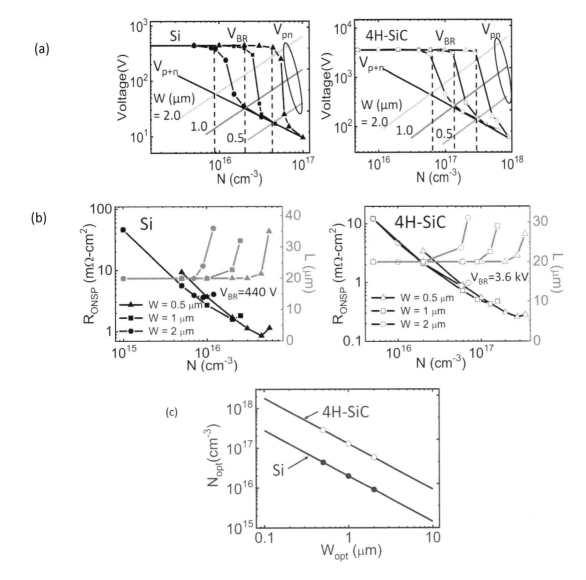

(a)

(b)

(c)

Fig. 3.2 Results for Si (low band gap material) and 4H-SiC (high band gap material). (a) Simulated doping dependence of SJ V_{BR}, and the voltages V_{pn} and V_{p+n} defined in the text. Dotted lines show optimum doping. (b) Doping dependence of the simulated pillar length and the specific on-resistance calculated using (3.10). (c) Optimum pillar doping for a given pillar width; line is as per (3.8) and points are from (b).

analytical integration, Fulop suggested to express α_{eff} in terms of a power law function of the breakdown field E_y along the critical path y as [12]

$$\alpha_{eff} = CE_y{}^g \tag{3.4}$$

The material specific constants C and g are obtained by fitting (3.4) to the Chynoweth's form of α_{eff} given by [13]

TABLE 3.1

Fitting constants of impact ionization coefficients as per Fulop (3.4) and Chynoweth (3.5)

Constant	Material			
	Si [11],[12]	4H-SiC [4]	GaN [14],[15]	Diamond [16]
A (cm^{-1})	9.00×10^5	6.85×10^6	1.50×10^5	9.44×10^4
B (Vcm^{-1})	1.80×10^6	1.41×10^7	1.41×10^7	1.90×10^7
C (V^{-g}cm^{g-1})	1.8×10^{-35}	4.0×10^{-48}	1.5×10^{-42}	8.0×10^{-18}
g	7	8	7	3

$$\alpha_{eff} = A \exp\left[-B/E_y\right]. \tag{3.5}$$

Literature has provided values of A and B for Si [12], c-axis of 4H-SiC [4], GaN [14],[15] and diamond [16]. The same literature also gives the values of C and g for all except diamond. We have determined the C and g for diamond in the present work, which reports all the values in Table 3.1 and the fits in Fig. 3.3 for ready reference.

Using (3.4) in (3.3), and assuming E_y to be a linear function of y, we can solve for y_d and then derive $V_{p+n} = qN y_d^2 / 2\varepsilon_s$ as

$$V_{p^+n} = \theta_V N^{-(g-1)/(g+1)} \tag{3.6}$$

where

$$\theta_V = 0.5\left(\frac{\varepsilon_s}{q}\right)^{(g-1)/(g+1)}\left(\frac{g+1}{C}\right)^{2/(g+1)} \tag{3.7}$$

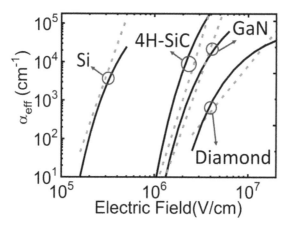

Fig. 3.3 Impact ionization coefficient vs electric field for Si, 4H-SiC, GaN and diamond in as per Chynoweth (3.5) (solid line) and Fulop (3.4) (dashed line).

Using (3.1) and (3.6) in (3.2), we derive the formula for N_{opt} as

$$N_{opt} = \theta_N W_{opt}^{-\left(1+g^{-1}\right)} \tag{3.8}$$

where

$$\theta_N = \left(\varepsilon_S \theta_V / q\right)^{0.5(1+g^{-1})} \tag{3.9}$$

Thus, N_{opt} is a function of W_{opt} alone for a given material unlike in the model presented for Si in Ref.[5] and its extended version for 4H-SiC in (A2), where it appears to depend on L_{opt} as well. Table 3.2 summarizes the values of θ_V and θ_N for various materials. The results of (3.8) shown in Fig. 3.2(c) for Si and 4H-SiC are found to be 30-36% higher than Ref. [5] and (A1) respectively.

To validate the above formula, we calculated the R_{ONSP} versus N curve for a fixed V_{BR} using

$$R_{ONSP} = 2LW / \left[qN\mu_n(W - W_d)\right], \tag{3.10}$$

where W_d is the zero bias depletion width of the junction between p and n pillars and μ_n is the electron mobility given by

$$\mu_n = \mu_{\min} + \frac{\mu_{\max} - \mu_{\min}}{1 + \left(N/N_0\right)^\gamma} \tag{3.11}$$

The values of μ_{max}, μ_{min}, N_0 and γ for Si [17], c-axis 4H-SiC [18], GaN [19] and diamond [20] are summarized in Table 3.3. Fig. 3.2(b) shows such R_{ONSP} versus N curves for a fixed $V_{BR} = 440$ V for Si and 3.6 kV for 4H-SiC; similar curves could be generated for GaN and diamond as well. For each N, the L was determined by TCAD. The N_{opt} is the location of minimum R_{ONSP} of this

TABLE 3.2

Values of constants in (3.7), (3.9) and (3.16) based on ionization coefficients of Table 3.1

Constant	Material			
	Si	4H-SiC	GaN	Diamond
θ_V (Vcm$^{-3(g-1)/(g+1)}$)	5.24×10^{13}	4.77×10^{15}	2.68×10^{15}	6.28×10^{11}
θ_N (cm$^{-2+1/g}$)	5.39×10^{11}	4.02×10^{12}	4.59×10^{12}	1.58×10^{12}
θ_L (cmV$^{-g/(g-1)}$)	1.92×10^{-6}	2.01×10^{-7}	1.27×10^{-7}	3.53×10^{-9}

TABLE 3.3

Parameters of the doping dependent mobility formula (3.11).

Constant	Material			
	Si [17]	4H-SiC [18]	GaN [19]	Diamond [20]
μ_{max} (cm^{2}V^{-1}s^{-1})	1417	950	1000	1030
μ_{min} (cm^{2}V^{-1}s^{-1})	52	40	55	0
N_0 ($\times 10^{16}$ cm^{-3})	9.68	20.00	20.00	9.90
γ	0.68	0.76	1.00	0.56

curve. Fig. 3.2(c) shows that the values of N_{opt} so obtained from Fig. 3.2(b) match with those of (3.8), thus validating (3.8). R_{ONSP} shoots up for $N > N_{opt}$ since V_{BR} starts falling in this regime (see Fig. 3.2(a)). To restore the V_{BR} to the fixed value (= 440 V for Si and 3.6 kV for 4H-SiC) for which the R_{ONSP} vs N is plotted, L needs to be increased (see Fig. 3.2(b). This increase in L is responsible for the rise in R_{ONSP}.

3.1.2 Optimum Pillar Length and Width

Denote the V_{BR} at $N = N_{opt}$ as $V_{BR,target}$. As shown in Fig. 3.2(a) and explained in the previous section, $V_{BRtarget}$ is somewhat lower than V_{SJ}. Extensive simulation of devices with several values of $V_{BR,target}$ and V_{SJ} for Si, SiC, GaN and diamond over $0.2 \leq W \leq 2$ μm yield

$$V_{BR,target} \approx 0.86\ V_{SJ} \tag{3.12}$$

within 8% error (see Fig. 3.4). The field distribution associated with V_{SJ}, which is V_{BR} for low N, turns out to be almost uniform over L, e.g. see E_y versus y for $N = 1 \times 10^{17}$ cm^3 in Fig. 3.1(a). Thus we have

$$E_y \approx V_{SJ}/L. \tag{3.13}$$

Substituting (3.13) in (3.4) and using the result in (3.3), performing the integration and rearranging, we get the pillar length L required to achieve a given V_{SJ} as

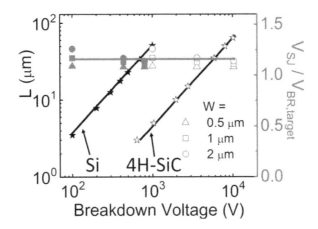

Fig. 3.4 Pillar length, L, for a given low doping breakdown voltage, V_{SJ}, as per (3.14), and the ratio of V_{SJ} to $V_{BR,target}$. Lines represent our model and points represent simulation results.

$$L = \left(C V_{SJ}{}^{g} \right)^{1/(g-1)}. \tag{3.14}$$

Fig. 3.4 shows that the results of (3.14) match with TCAD simulations. Substituting for V_{SJ} from (3.12) in (3.14) yields

$$L_{opt} = \theta_L \left(V_{BR,target} \right)^{g/(g-1)} \tag{3.15}$$

where

$$\theta_L = \left(1.16^g \, C \right)^{1/(g-1)}. \tag{3.16}$$

Thus, L_{opt} is a function of $V_{BR,target}$ alone for a given material unlike in the model

presented for Si in Ref.[5] and its extended version for 4H-SiC in (A2), where it appears to depend on W_{opt} as well. Table 3.2 summarizes the values of θ_L for various materials. Denoting the maximum aspect ratio limited by technology as r we get

$$W_{opt} = L_{opt} / 2r. \tag{3.17}$$

3.2 DESIGN EXAMPLE

Consider designing a 4H-SiC SJ with $V_{BR,target} = 3$ kV. Assume that the available ICP etching process [21] permits trenches with $r = 18$. Using the appropriate values of θ_V, θ_N, and θ_L from Table 3.2, and μ_{max}, μ_{min}, N_0 and γ from Table 3.3, (3.15) yields $L_{opt} = 18.5$ μm, (3.17) yields W_{opt} = 0.5 μm, (3.8) yields $N_{opt} = 2.8 \times 10^{17}$ cm^{-3} and (3.10) yields $R_{ONSP} = 0.22$ mΩ-cm^2. TCAD simulation with these L_{opt}, W_{opt}, N_{opt} yields V_{BR} = 3053 V within 2% of $V_{BR,target}$, validating our approach. The "extended" prior models (A1)-(A3) yield almost the same L_{opt} and W_{opt} but 36% lower N_{opt} = 1.8×10^{17} cm^{-3} and so 36% higher R_{ONSP}.

3.3 LOWER LIMIT OF SPECIFIC ON-RESISTANCE

Using the above procedure we calculated the R_{ONSP} over V_{BR} = 0.1-1 kV for Si devices with r = 12 [5] and over V_{BR} =1-10 kV for 4H-SiC devices with r = 5, 18 [21]. The results given in Fig. 3.5 show our R_{ONSP} to be 30-36 % lower than that obtained by the prior model for Si [5] and its "extended" version for 4H-SiC (see Appendix). These results confirm that our model yields the lower limit of R_{ONSP} for a given $V_{BR,target}$ and r allowed by technology.

Figure 3.6 compares the measured R_{ONSP} and V_{BR} of fabricated devices reported in literature with our model predictions. Charge imbalance due to poor process control is the cause of the higher R_{ONSP} of the fabricated devices. This is because, as pointed out in [26], one needs to use a doping which is much lower than N_{opt} to achieve the given $V_{BR,target}$ when charge imbalance is present. The ratio of the measured to modeled R_{ONSP} values indicated in the figure for each device highlights the scope for improving the process to reduce charge imbalance and parasitic resistances. This data should motivate technologists to research upon process improvement.

Although several technological challenges lie ahead in the fabrication of superjunction in GaN and diamond, it is of interest to investigate and compare their theoretical potential with

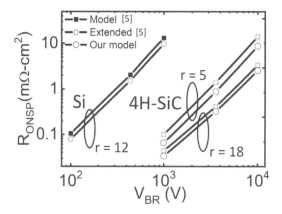

Fig. 3.5 Comparison of specific on-resistance R_{ONSP} versus breakdown voltage V_{BR} yielded by our model and the state of the art model for Si in [5] and its extended version for 4H-SiC in (A1)-(A3), for different aspect ratios, r.

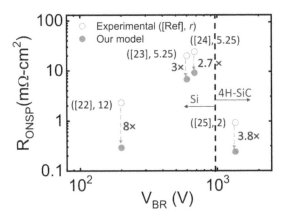

Fig. 3.6 Measured R_{ONSP} compared with the minimum R_{ONSP} predicted by our model for fabricated devices from literature. The aspect ratio, r, employed in fabrication and the ratio of the measured to minimum R_{ONSP} are also indicated.

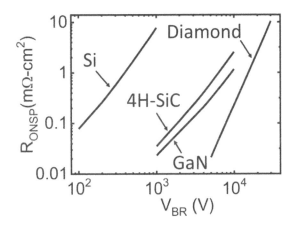

Fig. 3.7 Our model calculations of the specific on-resistance versus breakdown voltage of superjunctions with $r = 18$ in silicon, 4H-SiC, GaN and diamond.

Si and 4H-SiC counterparts. For illustration, we choose devices with $V_{BR} = 100$ V–1 kV in Si, 1–10 kV in 4H-SiC and GaN, and 5–30 kV in diamond. Fig. 3.7 shows the R_{ONSP} versus V_{BR} of these devices ignoring parasitic resistance and assuming $r = 18$. These calculations can be easily repeated for any other values of r and V_{BR}. By allowing such comparison of R_{ONSP} versus V_{BR} of several materials, our model allows determination of the application range of these materials.

3.4 SPECIFIC ON-RESISTANCE VERSUS PILLAR ASPECT RATIO

As described in Chapter 1, fabrication of an SJ often requires forming a trench for realizing p-pillars. It is of interest to study the behavior of R_{ONSP} with the aspect ratio of the p- pillar, r and determine the optimum value of r, denoted by r_0, that yields the minimum R_{ONSP} for a $V_{BR,target}$.

Analytical derivation of r_0 requires the solution of

$$dR_{ONSP}/dr\,|_{r=r_0} = 0$$ (3.18)

To derive an expression for r_0 using (3.18), we need R_{ONSP} as a function of r alone. By using N_d from (3.8) in (3.10), L from (3.15) in (3.10) and writing W_n in terms of r using (3.17), we get

$$R_{ONSP} = \left(\frac{\phi V_{BR}^{2+3(g-1)^{-1}}}{\varepsilon_S \mu_n} \right) \frac{r^{-\left(1+g^{-1}\right)}}{1 - r^{0.5\left(1-g^{-1}\right)}\sqrt{\lambda V_{bi}/V_{BR}}} .$$ (3.19)

$$\phi = 0.58\lambda\left[C^3 1.16^{(2g+1)} \right]^{1/(g-1)} \qquad \lambda = 2.44(g+1)^{-1/g} 2^{-0.5/g}$$

We solve for r_0 using the differential of the above in (3.18) and setting $V_{bi} \approx E_g/q$ (i.e. bandgap in Volts) [27]

$$r_0 = \left[1.16^g (g+1)^{2g+1}(3g+1)^{-2g} 2^{0.5g} \right]^{1/(g-1)} \left[V_{BR}/\left(E_g/q\right) \right]^{g/(g-1)}$$ (3.20)

3.4.1 Results and Discussion

For Si and 4H-SiC, we can simplify (3.19),(3.20) using $g \approx 7.5$ as

$$R_{ONSP} \approx \left(\frac{\phi V_{BR}^{2.46}}{\varepsilon_S \mu_n} \right) \frac{r^{-1.13}}{1 - r^{0.43}\sqrt{1.75 V_{bi}/V_{BR}}} \qquad r_0 \approx 0.235 \left(\frac{V_{BR}}{E_g/q} \right)^{1.15}$$ (3.21)

Fig. 3.8 plots (3.19), (3.20) for a 5 (0.5) kV balanced 4H-SiC (Si) SJ using the parameters of Table 3.1 and Table 3.3.

It is to be noted that the increase in r_0 with V_{BR} is superlinear (see (3.20) or (3.21)). For Si (4H-SiC) SJs with V_{BR} = 0.1–1 (1–10) kV, we obtain r_0 = 45–660 (165–2260) from (3.20). However, literature [28],[29] has reported pillar aspect ratios only up to 25 (3.5) in Si (4H-SiC) SJs. This creates an impression that there is significant scope in reduction of R_{ONSP} by realizing trenches with higher r by process improvement. However, it is to be noted that our analysis presented above is strictly valid only for a balanced SJ. Real devices are afflicted by process variations that introduce charge imbalance and must be considered for practical device design. This will be discussed in chapter 4.

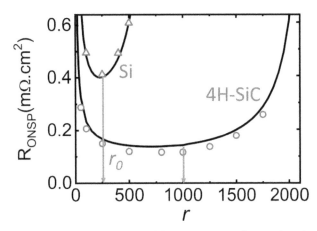

Fig. 3.8 Specific on-resistance, R_{ONSP}, versus pillar aspect ratio, r, for 4H-SiC (Si) SJ with V_{BR} = 5 (0.5) kV, using the parameters of Table 3.1 and Table 3.3. Points are TCAD simulation; lines are (3.19). Dashed (solid) lines correspond to 4H-SiC (Si) SJs.

3.4.2 Comparison with Prior Work

Ref. [27] estimated the U-shaped variation of R_{ONSP} with W_n of a balanced SJ, from which the minimum R_{ONSP} can be located at W_n = 0.09 (0.025) μm in Si (4H-SiC) SJs. Using these and $L \approx$ 4–61 (5–75) μm for V_{BR} = 0.1–1 (1–10) kV from Eq. (3.15) we obtain r_0 = 22–340 (100–1500) in contrast to our r_0 = 45–660 (165–2260) from (3.20). Our r_0 values yield W_n = 0.044 (0.017) μm from (3.17) that are 30-50% smaller than those in [27], the reason for which can be traced to our use of 30% higher N_d than in [27], that leads to lower W_d and in turn to lower W_n as per (3.10). Our optimum N_d corresponding to minimum R_{ONSP} is higher since it is estimated from (3.8) as per the improved model, while Ref. [27] estimates this N_d using an older analytical model of [3].

3.5 CONCLUSION

Based on new insights into the doping dependence of field distributions in a balanced SJ, we derived simple closed-form solutions for its optimum pillar doping, length and width for a target breakdown voltage. The solutions have a generic form which allows quick calculations of R_{ONSP} versus V_{BR} of superjunctions in a range of materials. They were validated with TCAD simulations. The solutions reveal that optimum length depends on breakdown voltage alone, and optimum doping on optimum width alone. Prior estimates of optimum doping were found to be lower and hence the specific on-resistance higher than actual by > 30%. Thus, an improved theoretical minimum of the specific on-resistance of an SJ has been established. Calculations of

R_{ONSP} versus V_{BR} based on our solution were presented for materials ranging from Si whose band gap is low to SiC, GaN and diamond whose band gaps are high. These calculations brought out the scope for minimizing the specific on-resistance of fabricated Si and SiC superjunctions (reported in literature) by process improvement. Our work should motivate process development for making superjunctions with lower specific on-resistance.

Appendix

The analytical solution of Ref.[5] was originally reported for Si material. Its extended version for 4H-SiC with material specific changes can be written as

$$N_{opt} = 2\varepsilon_S V_{BR,\text{target}} / \left[q L_{opt}^2 f(W_{opt}/L_{opt})\right] \quad \text{cm}^{-3} \tag{3.A1}$$

$$L_{opt}^7 = 4\times10^{-48}\left[1+37.65 f(W_{opt}/L_{opt})\right]V_{BR,\text{target}}^8 \quad \text{cm} \tag{3.A2}$$

$$f(t) = 1.697t - 0.7524t^2 - 1.9\times10^{-4}t^3 \tag{3.A3}$$

3.6 REFERENCES

[1] K. Akshay and S. Karmalkar, "Improved Theoretical Minimum of the Specific On-Resistance of a Superjunction," *Semicond. Sci. Technol.*, vol. 36, no. 1, p. 015021, Dec. 2020.

[2] K. Akshay and S. Karmalkar, "Optimum Aspect Ratio of Superjunction Pillars Considering Charge Imbalance", *IEEE Trans. Electron Devices*, vol. 68, no. 4, pp. 1798 - 1803, Apr. 2021.

[3] T. Fujihira, "Theory of semiconductor superjunction devices," *Jpn. J. Appl. Phys.*, vol. 36, no. 10, pp. 6254–6262, Oct. 1997, doi: 10.1143/JJAP.36.6254.

[4] L. Yu and K. Sheng. "Modeling and optimal device design for 4H-SiC super-junction devices," *IEEE Trans. Electron Devices*, vol. 55, no. 8, pp. 1961-1969, Jul. 2008, doi: 10.1109/TED.2008.926648.

[5] A. G. M. Strollo and E. Napoli, "Optimal on-resistance versus breakdown voltage tradeoff in superjunction power devices: A novel analytical model," *IEEE Trans. Electron Devices*, vol. 48, no. 9, pp. 2161–2167, Sept. 2001, doi: 10.1109/16.944211.

[6] X. B. Chen, P. A. Mawby, K. Board, and C. A. T. Salama, "Theory of a novel voltage-sustaining layer for power devices," *Microelectron. J.*, vol. 29, no. 12, pp. 1005–1011, Dec. 1998, doi: 10.1016/S0026-2692(98)00065-2.

[7] M. Alam, D. T. Morisette, and J. A. Cooper, "Design guidelines for superjunction devices in the presence of charge imbalance." *IEEE Trans. Electron Devices*, vol. 65, no. 8, pp. 3345-3351, 2018.

[8] B. Song, M. Zhu, Z. Hu, K. Nomoto, D. Jena and H. G. Xing, "Design and optimization of GaN lateral polarization-doped super-junction (LPSJ): An analytical study " in Proc. *Int. Symp. Power Semiconductor Device ICs*, pp. 125-128, 2015.

[9] ATLAS User's Manual, SILVACO International, 2006.

[10] J. Shields, "Breakdown in silicon pn junctions," *J. Electron. Control,* vol. 6, no. 2, pp. 130-148, Jan. 1959, doi: 10.1080/00207215908937136.

[11] J. Maserjian, "Determination of avalanche breakdown in pn junctions," *J. Appl. Phys.,* vol. 30, no. 10, pp. 1613-1614, 1959, doi: 10.1063/1.1735012.

[12] W. Fulop, "Calculation of avalanche breakdown voltages of silicon pn junctions," *Solid-St. Electron.,* vol. 10, no. 1, pp. 39-43, 1967, doi: 10.1016/0038-1101(67)90111-6.

[13] A. G. Chynoweth, "Ionization rates for electrons and holes in silicon", *Phys. Rev.,* vol. 109, p.1537, 1958, doi: 10.1103/PhysRev.109.1537.

[14] B. J. Baliga, "Gallium nitride devices for power electronic applications", *Semicond. Sci. Technol.,* vol. 28, no. 7, pp. 074011-1-074011-8, 2013. doi: 10.1088/0268-1242/28/7/074011.

[15] A. M. Ozbek, "Measurement of impact ionization coefficients in GaN," PhD Thesis, North Carolina State University, 2011.

[16] A. Hiraiwa and H. Kawarada, "Figure of merit of diamond power devices based on accurately estimated impact ionization processes", *J. Appl. Phys.,* vol. 114, no. 3, pp. 034506-1-034506-9, Jul. 2013, doi: 10.1063/1.4816312.

[17] B. Van Zeghbroeck, Principles of Semiconductor Devices. Englewood Cliffs, NJ, USA: Prentice-Hall, Dec. 2009.

[18] M. Roschke and F. Schwierz, "Electron mobility models for 4H, 6H, and 3C SiC," *IEEE Trans. Electron Devices,* vol. 48, no. 7, pp. 1442–1447, Jul. 2001, doi: 10.1109/16.930664.

[19] Z. Li and T. P. Chow, "Design and simulation of 5–20-kV GaN enhancement-mode vertical superjunction HEMT", *IEEE Trans. Electron Devices,* vol. 60, no. 10, pp. 3230-3237, Oct. 2013, doi: 10.1109/TED.2013.2266544.

[20] J. Pernot and S. Koizumi, "Electron mobility in phosphorous doped 111 homoepitaxial diamond", *Appl. Phys. Lett.,* vol. 93, no. 5, Aug. 2008, doi: 10.1063/1.2969066.

[21] K. M. Dowling, E. H. Ransom and D. G. Senesky, "Profile evalution of high aspect ratio silicon carbide trenches by inductive coupled plasma etching", *J. Microelectromech. Syst.,* vol. 26, no.1, pp. 135-142, Feb. 2017 doi: 10.1109/JMEMS.2016.2621131.

[22] Y. Hattori, K. Nakashima, M. Kuwahara, T. Yoshida, S. Yamauchi, and H. Yamaguchi, "Design of a 200V super junction MOSFET with n-buffer regions and its fabrication by trench filling," in *Proc. Int. Symp. Power Semiconductor Device ICs,* 2005, pp. 189–192, doi: 10.1109/WCT.2004.239903.

[23] W. Saito, L. Omura, S. Aida, S. Koduki, M. Izumisawa, H. Yoshioka, and T. Ogura, "A 20 mΩcm2 600 V-class superjunction MOSFET," in *Proc. Int. Symp. Power Semiconductor Device ICs,* 2004, pp. 459–462, doi: 10.1109/WCT.2004.239755.

[24] Y. Onishi, S. Iwamoto, T. Sato, T. Nagaoka, K. Ueno, and T. Fujihira, "24 mΩ.cm2 680 V silicon superjunction MOSFET," in *Proc. Int. Symp. Power Semiconductor Device ICs,* 2002, pp. 241–244, doi: 10.1109/ISPSD.2002.1016216.

[25] X. Zhong, B. Wang, J. Wang and K. Sheng, " Experimental demonstration and analysis of a 1.35-kV 0.92 mΩ.cm2 SiC superjunction Schottky diode", *IEEE Trans. Electron Devices*, vol. 65, no. 4, pp. 1458-1465, Apr. 2018, doi: 10.1109/TED.2018.2809475.

[26] K. Akshay and S. Karmalkar, "Quick Design of a Superjunction Considering Charge Imbalance Due to Process Variations", *IEEE Trans. Electron Devices*, vol. 67, no. 8, pp. 3024-3029, Aug. 2020, doi: 10.1109/TED.2020.2998443.

[27] H. Kang and F. Udrea, "True material limit of power devices—Applied to 2-D superjunction MOSFET," *IEEE Trans. Electron Devices*, vol. 65, no. 4, pp. 1432–1439, Apr. 2018.

[28] J. Sakakibara, Y. Noda, T. Shibata, S. Nogami, T. Yamaoka and H. Yamaguchi, "600 V-class super junction MOSFET with high aspect ratio P/N columns structure", *Proc. 20th Int. Symp. Power Semiconductor Device ICs*, pp. 299-302, May 2008, doi: 10.1109/ISPSD.2008.4538958.

[29] R. Kosugi, S. Ji, K. Mochizuki, K. Adachi, S. Segawa, Y. Kawada, et al., "Breaking the theoretical limit of 6.5 kV-class 4H-SiC super-junction (SJ) MOSFETs by trench-filling epitaxial growth", *Proc. 31st Int. Symp. Power Semicond. Devices ICs (ISPSD)*, pp. 39-42, May 2019. doi: 10.1109/ISPSD.2019.8757632.

CHAPTER 4

SUPERJUNCTION DESIGN IN THE PRESENCE OF CHARGE IMBALANCE

In Chapter 3, we discussed the modeling and design of an ideal balanced superjunction. However, a perfect charge balance between the n- and p- pillar cannot be achieved in practice due to the presence of inevitable process variations. The extent of such variations, dictated by the available technology, decides the amount of charge imbalance between the n- and p- pillar. It is found that the breakdown voltage, V_{BR}, of a superjunction whose pillar parameters, namely – doping, length and width, are optimized to yield the least specific on-resistance, R_{ONSP}, for the specified V_{BR}, falls drastically for even small charge imbalances. Thus, a structure fabricated with these target parameter values often has a V_{BR} much lower than specified due to the charge imbalance caused by random process variations.

In this chapter, we give generic closed form solutions valid across materials, for the alternate target parameter values which yield the specified V_{BR} in spite of process variations, with minimum sacrifice in R_{ONSP}. Our solution eliminates the tedious iterations of prior design methods. Further, we study the variation of R_{ONSP} with the pillar aspect ratio, r, and derive a simple closed form equation for its optimum value, r_0 as a function of V_{BR}. For practical Si and 4H-SiC SJs r_0 is found to be as low as 8–15 for V_{BR} = 0.1–10 kV and charge imbalance = 5–20 %. This is significantly different from the r_0 of a balanced superjunction which varies between 45–660 (165–2260) for Si (4H-SiC) SJs with V_{BR} = 0.1–1 (1–10) kV as shown in Chapter 3. This calls for caution while using balanced SJ theory to design practical SJs. This chapter is based on Ref. [1],[2].

In Section 4.1, the design approach is presented. In Section 4.2, the analytical solution for the optimum pillar doping, length and r_0 is derived, verified by TCAD simulation and compared with prior experimental and theoretical data by considering charge imbalance only due to imbalance in pillar doping. Section 4.3 discusses the modification required for the solution in

Section 4.2 when charge imbalance is due to imbalance in pillar width alone; Section 4.4 discusses the same when imbalance is due to both pillar doping and width.

4.1 DESIGN APPROACH

We introduce our approach with the help of the n-pillar doping dependence of TCAD simulated R_{ONSP} and V_{BR} of a typical superjunction as shown in Fig. 4.1. This data is based on the doping dependent mobility model reported in Ref.[3] and Selberherr's impact ionization model [4] with parameters of Ref. [5] whose results have been found to match with measurements by subsequent works [6], [7].

The R_{ONSP} curve of Fig. 4.1 corresponds to a balanced superjunction with $V_{BR}=3.6$ kV and is obtained using

$$R_{ONSP}= 2LW/\left[qN_d\mu_n\left(W-W_d\right)\right] \tag{4.1}$$

where μ_n is the electron mobility, W_d is the zero-bias n-pillar depletion width of the junction between p and n pillars given by

$$W_d \approx \sqrt{\varepsilon V_{bi}/qN_d} \ , \quad V_{bi} \approx 2V_t \ln(N_d/n_i) \tag{4.2}$$

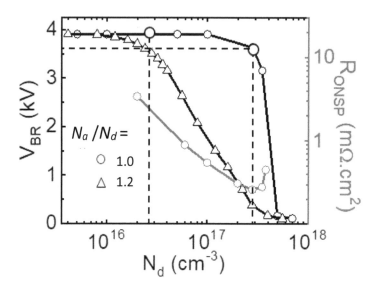

Fig. 4.1 TCAD simulated n-pillar doping dependencies of specific-on resistance, R_{ONSP}, and breakdown voltage, V_{BR}. The R_{ONSP} is estimated for a balanced superjunction with $V_{BR}=3.6$ kV; pillar length, L, is adjusted to maintain the V_{BR} constant. The V_{BR} curves are estimated for L, $W=$ 22, 0.5 μm which correspond to minimum R_{ONSP}.

where V_{bi} is the built-in voltage and n_i is the intrinsic concentration. The μ_n in (4.1) is estimated using the doping dependent mobility model given by

$$\mu_n = \mu_{min} + \frac{\mu_{max} - \mu_{min}}{1 + (N_d / N_0)^\gamma},$$

(4.3)

where μ_{max}, μ_{min}, N_0, and γ are material dependent fitting constants obtained by matching (4.3) with measured μ_n versus N_d data.

In (4.1), W is fixed at 0.5 μm corresponding to the least W considered in [8], [9] and L is adjusted for each N_d to maintain V_{BR} = 3.6 kV. The R_{ONSP} shoots up for N_d beyond the minimum point since V_{BR} starts falling in this regime. To restore V_{BR} to 3.6 kV, L needs to be increased which causes the rise in R_{ONSP}. The least R_{ONSP} is achieved at N_d = 2.8 x 10^{17} cm^{-3} and L = 22 μm. The V_{BR} curves of both balanced and imbalanced devices are generated using the above L and W values.

The imbalanced device considered has 20 % doping imbalance as per N_a = 1.2N_d. Note that a device with same N_d but lower N_a = N_d/1.2 also has 20% charge imbalance; however this device has a higher V_{BR}, since the V_{BR} is decided by the doping of the heavily doped pillar. We consider the structure with N_a = 1.2N_d which has the worst V_{BR} for the given charge imbalance.

Suppose, during fabrication, we target N_d = N_a = 2.8 x 10^{17} cm^{-3}, L = 22 μm and W = 0.5 μm which yield the least R_{ONSP}. Assume that process variations cause N_a to vary by ±15 % due to the difficulty in controlling the activation efficiency of p-dopants [8], but N_d to vary by ± 5 % due to the control achievable during epitaxial growth. In the worst case, the doping in the fabricated device would be either N_d = 1.2N_a or N_a = 1.2N_d which yields the worst V_{BR}. Hence, the worst V_{BR} will correspond to the imbalanced rather than the balanced curve of Fig. 4.1, falling to as low as 0.46 kV. Instead, we can ensure $V_{BR} \geq$ 3.6 kV in spite of the worst process variation if, during fabrication, we target a balanced device with about 10 times lower N_d = 2.7 x 10^{16} cm^{-3} at which the V_{BR} = 3.6 kV on the imbalanced curve. Thus, a balanced device whose L and W correspond to the least R_{ONSP} but the doping corresponds to an imbalanced device having the specified V_{BR} is immune to process variations. Below we derive closed-form solutions for these L, W and N_d = N_a that are within 5% of the TCAD simulated values for imbalance factors $\geq$ 10%. We also give an analytical solution which gives accurate values for imbalance factors down to 1% using a few iterations with the closed-form solution as the initial condition.

4.2 SUPERJUNCTION WITH DOPING IMBALANCE

Refer to Fig. 1.4(a). In this section, we treat the structure whose $W_n = W_p = W$, so that the charge imbalance is due to $N_a \neq N_d$ alone. Prior works [8]-[12] have defined the doping imbalance factor, k_N, in terms of the doping ratio N_a/N_d in different ways. Our work defines

$$
\begin{aligned}
k_N &= 1 - \left(N_a/N_d\right) \text{ for } N_d > N_a \\
&= 1 - \left(N_d/N_a\right) \text{ for } N_a > N_d
\end{aligned}
$$
(4.4)

As per this definition, if the doping levels of the two pillars are swapped, the value of k_N is unaltered just as that of V_{BR}. This allows a one to one map of k_N to V_{BR}. Without any loss of generality we assume $N_d > N_a$ as done in most of the prior works. However, the above definition of k_N allows the use of the equations of this section derived below for $N_a > N_d$ as well, by simply interchanging N_a and N_d.

4.2.1 Field Distribution at Breakdown

Refer to Fig. 4.2 showing the simulated vertical field distribution, E_Y, at breakdown at the edge of the n-pillar over L for various k_N. This distribution is exactly linear for $k_N = 1$, i.e. when

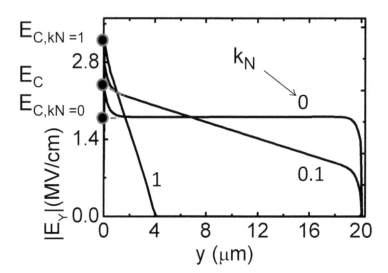

Fig. 4.2 Simulated vertical field distribution at breakdown over pillar length at $x = W$ (see Fig. 1.4 (a)) for different charge imbalance factors, k_N. Pillar parameters are $L = 20$ μm, $W = 0.56$ μm and $N = 8 \times 10^{16}$ cm^{-3}.

there is no p-pillar so that the device reduces to a simple p$^+$-n junction. The linearity is maintained over a significant portion of L even for $k_N < 1$. However, in accordance with Gauss's law, the slope reduces due to the depletion of the pillar charge by the lateral field, and becomes zero for $k_N = 0$ for which the device becomes balanced. Ref. [9] suggested that, for $k_N < 1$, we can regard the field to reduce linearly from an extrapolated peak value, E_C, with a slope = $q k_N N_d /2\varepsilon_S$. Based on this approximation, we write

$$V_{BR} \approx E_C L - 0.25\left(q k_N N_d / \varepsilon_S\right)L^2 .\tag{4.5}$$

4.2.2   Condition for Minimum Specific On-Resistance

Our design problem is to obtain the optimum pillar parameters, L_{opt}, W_{nopt} and N_{dopt} for which $V_{BR} = V_{BR,target}$ and R_{ONSP} is minimum, at specified k_N. According to the method of Lagrange multipliers [13], a function $f\,(a, b)$ is minimized under the constraint $g\,(a, b) = \lambda$, where λ is a constant, when a and b satisfy the condition $\partial_a f / \partial_b f = \partial_a g / \partial_b g$ (apart from $g\,(a, b) = \lambda$). In our case, $f \equiv R_{ONSP}$ given by (4.1), $g \equiv V_{BR}$ given by (4.5), $\lambda = V_{BRtarget}$, $a = L$ and $b = N_d$. Hence, R_{ONSP} is minimized under (4.5) with $V_{BR} = V_{BR,target}$ when L_{opt} and N_{dopt} satisfy

$$N_{dopt} = \frac{4\varepsilon_s V_{BR,t\arg et}}{q k_N L_{opt}^{\,2}}\left(\frac{2W_{nopt} - W_d}{4W_{nopt} - 3W_d}\right) \quad k_N >> 1/2r^2 .\tag{4.6}$$

This equation is not valid for $k_N \to 0$ for which it predicts an unrealistic value of $N_{dopt} \to \infty$. The lower limit of k_N for which this equation is valid is derived by considering the validity range of (4.5). Fig. 4.2 shows that for $k_N < 1$, the predominantly linear segment of the field distribution is preceded by a sharp peak, whose slope resembles that of the distribution for $k_N = 1$. This resemblance is because the peak occurs near the horizontal p$^+$-n junction (in the p-pillar, peak is near the n$^+$-p junction), where the pillar charge gets depleted mostly by the vertical rather than lateral field. Thinner this peak region better the validity of (4.5). This peak region is thin when the lateral depletion effect is strong, or equivalently, the applied voltage at which the pillars just get depleted by lateral one-dimensional field is $<< V_{BR}$, i.e.

$$\frac{q N_d W^2}{\varepsilon_S} << V_{BR} .\tag{4.7}$$

Although written for a balanced device, this equation proposed by us is a good approximation for imbalance levels, i.e. k_N, encountered in practice. Thus, (4.5) and hence (4.6) can be assumed to hold so long as (4.7) holds. Substitute for N_d in terms of k_N from (4.6) into (4.7). Recognize that the term in brackets on the LHS of (4.6) ranges from 1 for $W \to W_d$ to 1/2 for $W >> W_d$; use the large W limit of 1/2, since (4.7) restricts N_d to lower values for large W. Denoting the aspect ratio as

$$r = L/2W ,$$ (4.8)

we derive that (4.6) is valid for

$$k_N >> 1/2r^2 .$$ (4.9)

So far, r has varied in the range of 5–18.5 in fabricated SiC devices [14], [15], and 5–50 in modeling related works [8],[9]. Based on this data, the highest value of RHS in (4.9) is 0.02. Hence, (4.6) should work for all practical imbalance levels.

The N_{dopt}, L_{opt} and W_{nopt} values for a $V_{BR,target}$ can be obtained by solving (4.5), (4.6) and (4.8) provided we have analytical equations for E_C and r. We derive expressions for E_C using interpolation technique and for the optimum r, hereafter denoted as r_0, using minimization technique in the following sections.

4.2.3 Expression for E_C

Let E_y denote the breakdown field distribution along $x = W_n$ or $-W_p$ and

$$\alpha_{eff} = CE_y{}^g$$ (4.10)

denote Fulop's effective impact ionization coefficient where C and g are material dependent fitting constants obtained by matching (4.10) with experimental α_{eff} versus E_y data [16]. Table 4.1 gives the values of the constants in (4.3)-(4.10). We solve for E_C by setting the ionization integral, I, equal to unity,

$$I = \int_0^L \alpha_{eff}\, dy \to 1 .$$ (4.11)

The E_C can be obtained as an interpolating function between $E_{C,k_N=0}$, i.e. a balanced SJ and

TABLE 4.1
Values of constants in Eq. (4.3)-(4.10)

Constant	Material	
	Si [16],[17]	4H-SiC [8],[18]
μ_{max} (cm^{2}V^{-1}s^{-1})	1417	950
μ_{min} (cm^{2}V^{-1}s^{-1})	52	40
N_0 ($\times 10^{16}$ cm^{-3})	9.68	20.00
γ	0.68	0.76
g	7	8
C (V^{-g} cm^{g-1})	1.8×10^{-35}	4×10^{-48}
A_n, A_p (cm^{-1})	9.0×10^5	1.14×10^9, 6.85×10^6
B_n, B_p (V cm^{-1})	1.8×10^6	3.8×10^7, 1.41×10^7

$E_{C,k_N=1}$, i.e. an imbalanced SJ. For a balanced SJ, we approximate E_y to be uniform and equal to $E_{C,kN=0}$. Substituting for E_y in (4.10),(4.11) yields,

$$E_{C,k_N=0} = \left(CL\right)^{-1/g}.$$

(4.12)

The same approach can be applied to an imbalanced SJ. By substituting a linear E_y with a peak $E_{C,kN=1}$ and slope = qN_d/ε_S in (4.10) and (4.11), we get

$$E_{C,k_N=1} = \left[\left(\frac{qN_d}{\varepsilon_S}\right)\frac{g+1}{C}\right]^{1/(g+1)}.$$

(4.13)

The interpolating function is written as

$$E_C = (1 - k_N)^\beta E_{C,k_N=0} + k_N{}^\beta E_{C,k_N=1}$$

(4.14)

$$\beta \approx 0.8 + 1.85e^{-0.35r} \text{ for } r \geq 5,$$

where $E_{C,kN=0}$ and $E_{C,kN=1}$ are the extrapolated peak breakdown electric fields of balanced SJ and 1-D p$^+$-n junction respectively, and β is a fitting parameter.

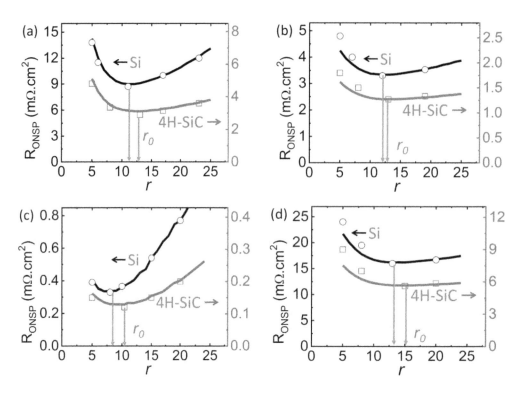

Fig. 4.3 Specific on-resistance, R_{ONSP}, versus pillar aspect ratio, r, for 4H-SiC (Si) SJ; (a) $V_{BR} = 5$ (0.5) kV, $k_N = 0.20$; (b) $V_{BR} = 5$ (0.5) kV, $k_N = 0.05$; (c) $V_{BR} = 1.0$ (0.1) kV, $k_N = 0.20$; (d) $V_{BR} = 10$ (1.0) kV, $k_N = 0.05$; lines are calculations using (4.1),(4.2) and (4.3) with pillar parameters from (4.6)-(4.15) and (4.12)-(4.14); points are TCAD simulations.

4.2.4 Expression for r_0

4.2.4.1 Specific On-Resistance versus Pillar Aspect Ratio Behavior

Prior to going through the intricacies of the analytical derivation of the optimum r (which minimizes R_{ONSP}), it is useful to gain some intuition regarding the shape of the R_{ONSP} versus r curve and the factors influencing it. For this purpose, we calculate L, W_n, and N_d iteratively for a given V_{BR}, k_N, r, and material using our MATLAB code [19], which works as follows. Start with an initial guess

$$L_{opt} \approx 1.06 L_{k_N = 0} \tag{4.15}$$

where $L_{k=0}$ was derived in Chapter 3 in the context of balanced SJs and is given by

$$L_{k_N=0} \approx \left\{ C \left(1.16 V_{BR,target} \right)^g \right\}^{1/(g-1)} \tag{4.16}$$

Next, W_{nopt} is evaluated from (4.8), N_{dopt} from (4.6), and a new estimate of L_{opt} is obtained by a

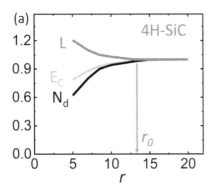
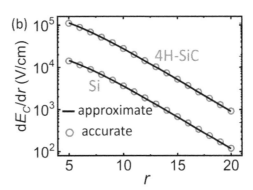

Fig. 4.4 (a) Normalized variations of pillar length, L, doping, N_d, and extrapolated critical field, E_C, with r; (b) dE_C/dr versus r; the curves correspond to the 5 (0.5) kV 4H-SiC (Si) SJ of Fig. 4.3(a). Similar variations are seen in other SJs.

quadratic solution of (4.5) with inputs from (4.12)-(4.14). The cycle (4.8),(4.6), (4.5) is repeated until convergence is obtained. Then we estimate R_{ONSP} from (4.1), (4.2), and (4.3). Such calculations for Si and 4H-SiC SJs in the practical range of $V_{BR,target}$ and k_N values lead to a U-shaped R_{ONSP} versus r behavior shown in Fig. 4.3, whose salient features are explained below.

Factors contributing to the U-shape can be understood with the help of Fig. 4.4(a) which shows the qualitative variations in N_d, L, and E_C as r is raised by shrinking W_n. For $r < r_0$, we have $W_n >> W_d$ and hence the variation in W_n does not affect N_d as per (4.6); however, β falls thereby increasing E_C as per (4.14), which causes L to fall to maintain the V_{BR} constant as per (4.5), which in turn causes N_d to rise as per (4.6); the fall in L and rise in N_d both reduce R_{ONSP} as per (4.1). For $r > r_0$, the β saturates and so E_C, L, and N_d saturate too; however, W_n approaches W_d shrinking the conduction area and raising R_{ONSP}. The results for typical Si and 4H-SiC SJs given in Fig. 4.3(a),(b) show that the r_0 as well as the flatness of the minimum increase as either V_{BR} increases or k_N falls. Fig. 4.3(c) and Fig. 4.3(d) show the curves for the extreme combinations of low V_{BR}–high k_N and high V_{BR}–low k_N, respectively, to bring out the variation in the r_0 and flatness of the minimum over the entire practical range of V_{BR} and k_N values. The r_0 is seen to vary in the range 8–13 for Si and 10–15 for 4H-SiC SJs.

The next section presents an analytical model of R_{ONSP} versus r which captures the above features and gives a closed-form solution for r_0.

4.2.4.2 Analytical Model for Imbalanced Superjunctions

Analytical derivation of r_0 requires the solution of

$$dR_{ONSP}/dr\,|_{r=r_0} = 0.$$

(4.17)

As a first step to expressing R_{ONSP} in terms of r for this purpose, in (4.1), we express W_n in terms of L and r to make R_{ONSP} a function of L, N_d, and r. However, expressing dL/dr and dN_d/dr in terms of r is difficult, while as we will show shortly, dE_C/dr can be expressed in terms of r by a simplification without sacrificing accuracy. Hence, we express L and N_d in terms of E_C as follows. An SJ would be designed to have $r = r_0$ to achieve minimum R_{ONSP}. Using the values of r_0, N_d and L estimated as per the previous section in (4.2), (4.8), we find that SJs have $W_n \gg W_d$ for values of r well beyond r_0. Hence, assuming $W_n \gg W_d$ in (4.6) and substituting the result into (4.5), we get

$$L = 3V_{BR} / 2E_C,$$

(4.18)

whose substitution back into (4.6) gives

$$N_d = 8\varepsilon_S E_C^2 / 9qk_N V_{BR} \qquad k_N \gg 1/2r^2.$$

(4.19)

Using the above L and N_d, and (4.8) for W_n in (4.1) we get R_{ONSP} as a function of E_C and r as

$$R_{ONSP} = \left(\frac{27k_N V_{BR}^2}{4\varepsilon_S \mu_n} \right) \frac{E_C^{-3}}{1 - r\sqrt{2k_N V_{bi}/V_{BR}}} \qquad k_N \gg 1/2r^2,$$

(4.20)

This equation captures the features of R_{ONSP} versus r behavior mentioned in the previous section. For small r, the decrease in E_C^{-3} with r (see Fig. 4.4(a)) dominates over the decrease in the denominator term leading to a decrease in R_{ONSP}; for large r, the E_C^{-3} saturates while the denominator term continues to fall leading to an increase in R_{ONSP}, and thereby, a U-shaped R_{ONSP} versus r curve. Substituting (4.20) into (4.17), we get

$$\frac{E_C}{3} \left(\frac{dE_C}{dr} \right)^{-1} = \sqrt{\frac{V_{BR}}{2k_N V_{bi}}} - r_0.$$

(4.21)

Differentiation of (4.14) considering the r dependence of β, but setting L in (4.12) and N_d in (4.13) as constants equal to their saturation values, (see Fig. 4.4(a)) leads to

$$\frac{dE_C}{dr} \approx 0.65 e^{-0.35r} \left[\begin{array}{c} E_{C,kN=0}(1-k_N)^\beta \log(1-k_N) \\ +E_{C,kN=1} k_N^\beta \log k_N \end{array} \right].$$

(4.22)

Fig. 4.4(b) shows that values of (4.22) match the accurate dE_C/dr obtained using iterative

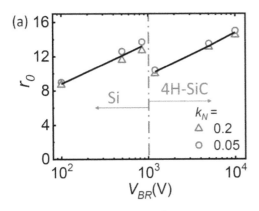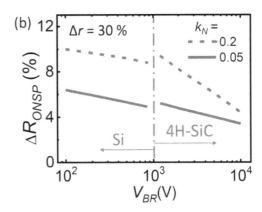

Fig. 4.5 (a) Optimum aspect ratio, r_0, versus V_{BR} in Si and 4H-SiC SJs. Points represent TCAD simulations for typical charge imbalance levels. The solid line represents our model (4.23). (b) Percentage change in R_{ONSP} due to ± 30% change in aspect ratio r around r_0 estimated using (4.23).

numerical calculations of the previous section which throw up r dependent N_d and L as in Fig. 4.4(a).

To get an approximate closed-form solution for r_0 we neglect r_0 compared to the other terms of (4.21) based on calculations of the terms for typical V_{BR} and k_N. Next, we substitute (4.22) into LHS of (4.21) and set $\beta \approx 0.8$ for $r \geq 8$. Further, we set $E_{C,kN=1} \approx 2E_{C,kN=0}$ [20] and $V_{bi} \approx E_g/q$ (i.e. bandgap in Volts) [21] as done in the design of a balanced SJ to get

$$r_0 \approx 1 + \frac{10}{7}\left[\ln\left(\frac{V_{BR}}{E_g/q}\right) + \ln\left\{\frac{-\ln\left[(1-k_N)k_N^{\ m}\right]}{(1+m)\sqrt{k_N}}\right\}\right] \approx 3 + \frac{10}{7}\ln\left(\frac{V_{BR}}{E_g/q}\right)$$

$$m = 2\left(k_N/\left[1-k_N\right]\right)^{0.8} \qquad k_N \gg 1/2r^2, \tag{4.23}$$

showing that r_0 rises logarithmically with $V_{BR}/(E_g/q)$. The simplified expression is obtained by approximating the complex k_N-dependent term (= 1.7, 1.1 for k_N = 0.05, 0.20) by 1.4, and predicts r_0 with < 5% error. The results are shown in Fig. 4.5(a) and agree with r_0 = 8–15 determined in section 4.2.4.1.

We examined the increase in R_{ONSP} from its minimum value due to the use of an approximate rather than accurate value of r_0. Calculations based on (4.20) given in Fig. 4.5(b) show that even a ± 30 % change in r around r_0 increases R_{ONSP} by ≤ 10%; the increase is lower for lower k_N and higher V_{BR}. What this "flat" minimum of R_{ONSP} versus r data implies is that even if the superjunction were designed with an approximate r_0 that differs from the accurate r_0 by ± 30 %, the R_{ONSP} would remain within 10% of the minimum, so long as the N_d, L, W_n are estimated corresponding to this approximate r_0 as suggested in the first paragraph of section 4.2.4.1. Hence, a superjunction designed quickly using the simple r_0 formula (4.23) would be as optimum as that using detailed time consuming TCAD simulations. This leads to the following quick design procedure based on closed-form calculations.

4.2.5 Closed-Form Design Equations

Once the value of r is restricted to 8–15, the accurate iterative solution of L_{opt}, W_{nopt}, and N_{dopt} for a $V_{BR,target}$ in section 4.2.4.1 can be reduced to closed-form as follows. Accurate values of L_{opt} fit into

$$L_{opt} \approx 1.16 L_{k_N = 0} \qquad (4.24)$$

where $L_{k=0}$ is given by (4.16). The values of L_{opt} given by (4.24) is within 5 % error for $k_N \leq 0.2$ and Si (4H-SiC) SJs with $V_{BR,target}$ = 0.1–1 (1–10) kV. Next, W_{nopt} is obtained using (4.23) in (4.8), and   is calculated from (4.6) where W_d is obtained from (4.2) using the value of N_d with $W_d = 0$. Fig. 4.6 gives the values of pillar parameters and Fig. 4.7 the R_{ONSP} calculated from these.

4.2.6 Model Validation

We compare the results of our model with TCAD simulations [4] which employ the SELB impact ionization model given by the following equation during reverse bias,

$$\alpha_{n,p} = A_{n,p} \exp\left(- B_{n,p} / E_y\right), \qquad (4.25)$$

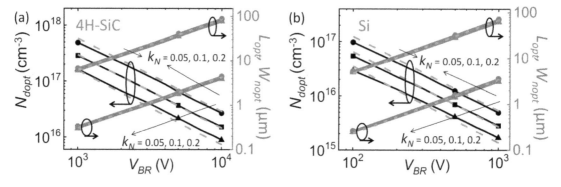

Fig. 4.6 The pillar parameters versus V_{BR} for 4H-SiC **(a)** and Si **(b)** SJ. Legend: Solid (dashed) lines - iterative (closed-form) solutions; points – TCAD.

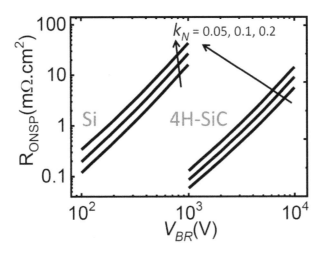

Fig. 4.7 Minimum specific on-resistance, R_{ONSP}, versus breakdown voltage, V_{BR}, in the presence of charge imbalance, k_N. The curves are generated using (4.1) with L, W_n, and N_d calculated using our analytical solution given in Section 4.2.4.1.

where α_n (α_p) is the impact ionization coefficient of electron (hole). A_n and B_n (A_p and B_p) are material dependent fitting constants obtained by matching (4.25) with experimental α_n (α_p) versus E_y data. The simulations also employ the ANALYTIC CONMOB concentration dependent mobility model given by (4.3) during forward bias, using the parameters listed in Table 4.1 which have been calibrated against the measured data summarized in Ref. [8], [16], [18]-[22]. Note that, as reported in literature, (4.10) is an empirical fit to (4.25), with $\alpha = \alpha_n = \alpha_p$ for Si [22] where both electron and hole ionization are equally significant, and $\alpha = \alpha_p$ for 4H-SiC [8],[23] where electron ionization is negligible. In on-state simulations, the p$^+$ and n$^+$ layers in Fig. 1.4(a) were replaced by ohmic contacts, as suggested in [8]. This is to avoid conductivity modulation resulting from bipolar injection, an effect extrinsic to superjunction structures and absent when such structures are employed as drift layer in unipolar devices.

The above TCAD simulations agree with our model calculations. Fig. 4.8 shows this agreement for the basic equation (4.14) for E_C varying with r, V_{BR} and k_N in range $r \geq 7$ where the r_0 values lie. Considering imbalanced SJs, Fig. 4.3 establishes this agreement for R_{ONSP} versus r around r_0 based on either (4.20) or equations (4.1) - (4.14), Fig. 4.5(a) for r_0 using (4.23), and Fig. 4.6 for the closed-form design equations of section 4.2.5.

Consider the validity condition $k_N \gg 1/2r^2$ for (4.6). A detailed comparison of analytical and TCAD simulated R_{ONSP} versus r results such as the one in Fig. 4.3 suggests that these R_{ONSP}

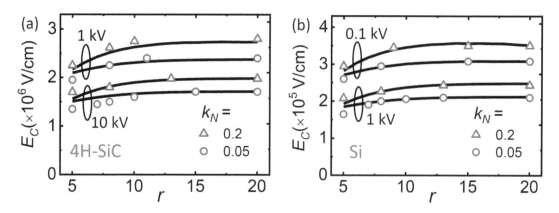

Fig. 4.8 Critical electric field, E_c, versus pillar aspect ratio, r, in 4H-SiC **(a)** and Si **(b)** SJs having widely different breakdown voltage, V_{BR}, and charge imbalance, k_N. Legend: solid lines – Eq. (4.14); points – TCAD.

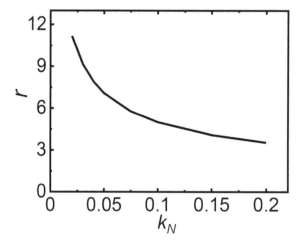

Fig. 4.9 A plot of $r = \sqrt{2.5/k_N}$. Our model and design procedure are valid for values of r above this curve, where their R_{ONSP} predictions match with TCAD simulations within 5 % error.

values differ by ≤ 5 % for $k_N \geq 5 \times (1/2r^2) = 2.5/r^2$ or equivalently $r \geq \sqrt{2.5/k}$, which therefore becomes the region of validity of our model and design procedure. The pictorial illustration of this validity region in Fig. 4.9 facilitates the confirmation of the validity of our approach for a given structure at hand. For instance, it makes it immediately clear why the analytical results and TCAD points match for values of r down to 5 in Fig. 4.3(a),(c) where $k_N = 0.2$, but diverge for $r < 7.1$ in Fig. 4.3(b),(d) where $k_N = 0.05$, and match around $r_0 = 8 - 15$ in all cases of Fig. 4.3 covering the range $k_N = 0.05 - 0.2$. Stated differently, for typical Si (4H-SiC) SJs, we have $r_0 \approx 10$ (12) and so our design approach of section 4.2.5 works for $k_N > 2.5$ (1.7) %.

4.2.7 Comparisons

4.2.7.1 Experimental

Consider design of a Si SJ with $V_{BR} = 600$ V. Ref. [24] fabricated SJs with $L = 37$ μm, average $r = 25$ or $W_n = 0.75$ μm and $N_d = 1 \times 10^{16}$ cm^{-3}. Fitting TCAD simulation into measured V_{BR} values for different charge imbalance levels, Ref. [24] extracted a charge imbalance of 5% corresponding to $V_{BR} = 600$ V. Practical SJs have edge terminations to prevent degradation of the V_{BR} below the ideal value predicted for plane junctions employed in simulations. In the absence of information on the edge termination structure in [24], we assume that the measured V_{BR} was close to the ideal simulated value. Under this assumption, our design equations suggest that, for 5% charge imbalance, i.e. $k_N = 0.05$, we can get $V_{BR} = 600$ V using the same L and N_d but a much lower and comfortable trench aspect ratio of $r_0 = 12$ or a thicker pillar of $W_n = 1.5$ μm and consequently 10% lower R_{ONSP}.

4.2.7.2 Balanced Superjunction

The theory and design of a balanced SJ, i.e. one with $k_N = 0$, was discussed in Chapter 3. It was shown that the minimum R_{ONSP} and r_0 are given by

$$R_{ONSP} = \left(\frac{\phi V_{BR}^{2+3(g-1)^{-1}}}{\varepsilon_S \mu_n} \right) \frac{r^{-\left(1+g^{-1}\right)}}{1 - r^{0.5\left(1-g^{-1}\right)}\sqrt{\lambda V_{bi}/V_{BR}}}. \tag{4.26}$$

$$\phi = 0.58\lambda\left[C^3 1.16^{(2g+1)}\right]^{1/(g-1)} \quad \lambda = 2.44(g+1)^{-1/g}2^{-0.5/g}$$

and

$$r_0 = \left[1.16^g (g+1)^{2g+1}(3g+1)^{-2g}2^{0.5g}\right]^{1/(g-1)}\left[V_{BR}/\left(E_g/q\right)\right]^{g/(g-1)}. \tag{4.27}$$

Further, it was shown that for Si and 4H-SiC, we can simplify (4.26),(4.27) using $g \approx 7.5$ as

$$R_{ONSP} \approx \left(\frac{\phi V_{BR}^{2.46}}{\varepsilon_S \mu_n} \right) \frac{r^{-1.13}}{1 - r^{0.43}\sqrt{1.75V_{bi}/V_{BR}}} \quad r_0 \approx 0.235\left(\frac{V_{BR}}{E_g/q} \right)^{1.15} \tag{4.28}$$

Fig. 4.10 plots the comparison of the R_{ONSP} versus r for balanced and imbalanced 4H-SiC (Si) SJs with $V_{BR} = 5$ (0.5) kV, using the parameters of Table 4.1. It can be seen that the r_0 of balanced devices are several tens of times more than those of imbalanced devices. Moreover, the increase in r_0 with V_{BR} is superlinear in a balanced SJ (see (4.27) or (4.28)) but logarithmic in an

66

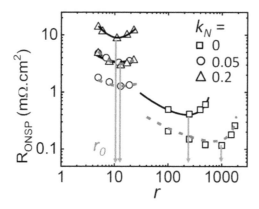

Fig. 4.10 Comparison of the specific on-resistance, R_{ONSP}, versus pillar aspect ratio, r, for balanced and imbalanced 4H-SiC (Si) SJs with $V_{BR} = 5$ (0.5) kV, using the parameters of Table 4.1. Points are TCAD simulations. Dashed (solid) lines represents 4H-SiC (Si) SJs.

imbalanced SJ (see (4.23)). Thus, it is inappropriate to use r_0 of balanced SJs to design practical SJs afflicted by charge imbalance from process variation.

The difference in r_0 of balanced and imbalanced devices can be better understood with the help of the simplified equation (4.28), whose comparison with (4.20) of the imbalanced case shows the following. Firstly, E_C^{-3} in (4.20) decreases rapidly and saturates around $r = 8{-}15$ (see Fig. 4.4(a)) while the decrease in $r^{-1.3}$ in (4.28) is slower but continues until large r. Secondly, the denominator of (4.20) decreases more rapidly/ than that of (4.28) as r increases more rapidly than $r^{0.43}$. These two factors cause R_{ONSP} for an imbalanced SJ to rise for lower values of r than in a balanced SJ.

4.3 SUPERJUNCTION WITH WIDTH IMBALANCE

In this section, we treat the structure whose $N_d = N_a$, so that the charge imbalance is due to $W_n \neq W_p$ alone. We define the width imbalance factor, k_W, in analogy to (4.4) for k_N, i.e.

$$k_W = 1 - \left(W_p/W_n\right) \text{ for } W_n > W_p$$
$$= 1 - \left(W_n/W_p\right) \text{ for } W_p > W_n. \tag{4.29}$$

and assume $W_n > W_p$ without any loss of generality. It is the imbalance in *charge* between the p- and n- pillars that degrades the device performance compared to a balanced superjunction. As pillar charge = pillar width × pillar doping, the doping or geometry have similar effect on the performance of a superjunction [9]. Specifically, a given k_W has the same effect on V_{BR} as a k_N whose value equals k_W.

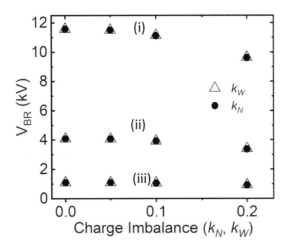

Fig. 4.11 Sensitivity of breakdown voltage to charge imbalance. For zero charge imbalance, (i) L = 74 μm, $N_a = N_d = 9.6 \times 10^{15}$ cm^{-3}, $W_n = W_p = 2$ μm. (ii) $L = 23$ μm, $N_a = N_d = 3.6 \times 10^{16}$ cm^{-3}, $W_n = W_p = 0.5$ μm. (iii) $L = 5.2$ μm, $N_a = N_d = 2 \times 10^{17}$ cm^{-3}, $W_n = W_p = 0.05$ μm.

Fig. 4.11 reports TCAD simulations illustrating this principle, which allows us to treat the width imbalance problem as an equivalent doping imbalance problem. Thus all our equations of the previous section apply after replacing k_N by k_W and W by W_n.

If $W_p > W_n$, all the equations of this section apply after interchanging W_p and W_n.

4.4 SUPERJUNCTION WITH DOPING AND WIDTH IMBALANCE

The results of the previous two sections allow us to treat a superjunction with both doping and charge imbalance as one with an equivalent doping imbalance. In analogy to (4.4) and (4.29), we can define charge imbalance factor as $k_{eff} = 1- (N_a W_p / N_d W_n)$ for $N_a W_p > N_d W_n$ and $k_{eff} = 1- (N_d W_n / N_a W_p)$ for $N_d W_n > N_a W_p$. This can be succinctly written as

$$k_{eff} = 1 - \left(1 - k_N\right)\left(1 - k_W\right) \tag{4.30}$$

where k_N is given by (4.4) and k_W by (4.29). All our equations of above sections apply after replacing k_N by k_{eff} and W by W_n. We assume $N_d W_n > N_a W_p$ without any loss of generality. If $N_a W_p > N_d W_n$, the above equations will apply after interchanging N_d with N_a, and W_n with W_p.

4.5 CONCLUSION

We gave a simple and generic closed form solution for the optimum pillar parameters of a superjunction which ensure the specified breakdown voltage in the presence of charge imbalance caused by random process variations, with minimum sacrifice in the specific on-resistance. The solution eliminates tedious iterations involved in prior design methods. It shows that, as compared to a balanced device with minimum specific on-resistance, the device optimized considering typical charge imbalance has significantly lower pillar doping but almost the same pillar length and width. Presently, it is believed that R_{ONSP} can be progressively reduced by raising r, and so literature has attempted pillar aspect ratios > 25 (10) in Si (4H-SiC) SJs. Counterintuitively, we showed that the optimum pillar aspect ratio is as low as 8–15 for 0.1–10 kV Si and 4H-SiC imbalanced SJs and increases logarithmically with the (breakdown voltage / material bandgap) ratio. On the other hand, the optimum aspect ratio for balanced SJs is several times higher and rises superlinearly with the (breakdown voltage/material bandgap) ratio, calling for caution while using balanced SJ theory to design practical SJs.

4.6 REFERENCES

[1] K. Akshay and S. Karmalkar, "Optimum Aspect Ratio of Superjunction Pillars Considering Charge Imbalance", *IEEE Trans. Electron Devices*, vol. 68, no. 4, pp. 1798 - 1803, Apr. 2021.

[2] K. Akshay and S. Karmalkar, "Quick Design of a Superjunction Considering Charge Imbalance Due to Process Variations," *IEEE Trans. Electron Devices*, vol. 67, no. 8, pp. 3024 3029, Aug. 2020.

[3] M. Roschke and F. Schwierz, "Electron mobility models for 4H, 6H, and 3C SiC," *IEEE Trans. Electron Devices*, vol. 48, no. 7, pp. 1442–1447, Jul. 2001.

[4] ATLAS User's Manual, SILVACO International, 2006.

[5] A. O. Konstantinov, Q. Wahab, N. Nordell, and U. Lindefeltd, "Ionization rates and critical fields in 4H silicon carbide," *Appl. Phys. Lett.*, vol. 71, no. 1, pp. 90–92, Mar. 1997, doi: 10.1063/1.119478.

[6] X. Li, "Design and simulation of high voltage 4H silicon carbide power devices," Ph.D. dissertation, Rutgers Univ., Piscataway, NJ, USA, 2005.

[7] X. Zhong, B. Wang and K. Sheng, "Experimental Demonstration and Analysis of a 1.35-kV 0.92-mΩ.cm^2 SiC Superjunction Schottky Diode." *IEEE Trans. Electron Devices*, vol. 65, no. 4, pp. 1458-1465, Apr. 2018, doi: 10.1109/TED.2018.2809475.

[8] L. Yu and K. Sheng. "Modeling and optimal device design for 4H-SiC super-junction devices," *IEEE Trans. Electron Devices*, vol. 55, no. 8, pp. 1961-1969, Jul. 2008, doi: 10.1109/TED.2008.926648.

[9] M. Alam, D. T. Morisette, and J. A. Cooper, "Design guidelines for superjunction devices in the presence of charge imbalance." *IEEE Trans. Electron Devices*, vol. 65, no. 8, pp. 3345-3351, Aug. 2018, doi: 10.1109/TED.2018.2848584.

[10] E. Napoli, H.Wang, and F. Udrea, "The effect of charge imbalance on superjunction power devices: An exact analytical solution," *IEEE Electron Device Lett.*, vol. 29, no. 3, pp. 249–251, Mar. 2008, doi: 10.1109/LED.2007. 915375.

[11] H. Wang, E. Napoli, and F. Udrea, "Breakdown voltage for superjunction power devices with charge imbalance: An analytical model valid for both punch through and non punch through devices," *IEEE Trans. Electron Devices*, vol. 56, no. 12, pp. 3175–3183, Dec. 2009, doi: 10.1109/TED.2009. 2032595.

[12] M. Alam, D. T. Morisette, and J. A. Cooper, "Practical design of 4H-SiC superjunction devices in the presence of charge imbalance," in *Proc. Int. Conf. Silicon Carbide Related Mater.*, 2017, pp. 563–567, doi: MSF.924.563.

[13] D. P. Bertsekas, Constrained Optimization and Lagrange Multiplier Methods, Academic Press, 1982.

[14] R. Kosugi, Y. Sakuma, K. Kojima, S. Itoh, A. Nagata, T. Yatsuo, Y. Tanaka and H. Okumura, "Development of SiC super-junction (SJ) device by deep trench-filling epitaxial growth," *Mater. Sci. Forum*, vols. 740–742, pp. 785–788, 2013, doi: MSF.740-742.785.

[15] K. M. Dowling, E. H. Ransom and D. G. Senesky, "Profile evalution of high aspect ratio silicon carbide trenches by inductive coupled plasma etching", *J. Microelectromech. Syst.*, vol. 26, no.1, pp. 135-142, Feb. 2017, doi: 10.1109/JMEMS.2016.2621131.

[16] W. Fulop, "Calculation of avalanche breakdown voltages of silicon pn junctions," *Solid-St. Electron.*, vol. 10, no. 1, pp. 39-43, 1967, doi: 10.1016/0038-1101(67)90111-6.

[17] B. Van Zeghbroeck, Principles of Semiconductor Devices. Englewood Cliffs, NJ, USA: Prentice-Hall, Dec. 2009.

[18] M. Roschke and F. Schwierz, "Electron mobility models for 4H, 6H, and 3C SiC," *IEEE Trans. Electron Devices*, vol. 48, no. 7, pp. 1442–1447, Jul. 2001, doi: 10.1109/16.930664.

[19] K. Akshay and S. Karmalkar, "Superjunction design using a bathtub curve." Accessed: Oct. 02, 2020. [Online]. Available: https://codeocean.com/capsule/0023716.

[20] A. G. M. Strollo and E. Napoli, "Optimal on-resistance versus breakdown voltage tradeoff in superjunction power devices: A novel analytical model," *IEEE Trans. Electron Devices*, vol. 48, no. 9, pp. 2161–2167, Sept. 2001, doi: 10.1109/16.944211.

[21] H. Kang and F. Udrea, "True material limit of power devices—Applied to 2-D superjunction MOSFET," *IEEE Trans. Electron Devices*, vol. 65, no. 4, pp. 1432–1439, Apr. 2018, doi: 10.1109/TED.2018.2808181.

[22] J. Maserjian, "Determination of avalanche breakdown in pn junctions," *J. Appl. Phys.*, vol. 30, no. 10, pp. 1613-1614, 1959, doi: 10.1063/1.1735012.

[23] R. Raghunathan and B. J. Baliga, "Temperature dependence of hole impact ionization coefficients in 4H and 6H-SiC", *Solid State Electron.*, vol. 43, no. 2, pp. 199-211, Feb. 1999, doi: https://doi.org/10.1016/S0038-1101(98)00248-2.

[24] J. Sakakibara, Y. Noda, T. Shibata, S. Nogami, T. Yamaoka and H. Yamaguchi, "600 V-class super junction MOSFET with high aspect ratio P/N columns structure", *Proc. 20th ISPSD*, pp. 299-302, May 2008, doi: 10.1109/ISPSD.2008.4538958.

CHAPTER 5

CHARGE SHEET SUPERJUNCTION IN 4H-SiC

SuperJunction (SJ) in silicon (Si) is commercially manufactured as its fabrication technology is sufficiently matured. However, fabrication of SJ in Silicon Carbide (SiC) material has several technological hurdles related to the realization of p-pillars, which need to be solved before it can be produced at a commercial scale reliably and profitably. One of the key hurdles is the poor control over the p- dopant activation efficiency in SiC. This makes the p- pillar charge difficult to control which consequently makes the device prone to severe charge imbalance. As increase in charge imbalance degrades the breakdown voltage V_{BR}, the device may fall short of target V_{BR}, $V_{BR,target}$, if charge imbalance increases beyond the designed tolerable limits.

We propose that these difficulties related to the p-pillar in SiC material be overcome using a version of SJ called Charge Sheet SJ (CSSJ) which we proposed in the context of Si material a decade ago [1], [2]. In CSSJ, the p-pillar is replaced by Al_2O_3 which has a fixed negative charge at its interface with the SiO_2 used as a liner before depositing Al_2O_3; this negative charge simulates the ionized p-dopants, and can be easily controlled via the Al_2O_3 deposition temperature. The cross section of the unit cell of an SJ and CSSJ are shown in Fig. 5.1(a) and Fig. 5.1(b) respectively.

In this chapter, we build the motivation for fabricating the CSSJ in SiC by discussing its operation, practicability, V_{BR} model, and design equations validated by TCAD simulations for V_{BR} in the range 1 to 10 kV and ≤ 20 % charge imbalance. This chapter is based on Ref. [3],[4],[5].

Section 5.1 discusses the device operation in both on- and off-state. Section 5.2 discusses the practicability aspects by outlining the potential fabrication steps of CSSJ and by explaining its superiority over the SJ fabrication steps in SiC material. A model for the V_{BR} of CSSJ as a function of device parameters is derived in Section 5.3 and the detailed device design steps are discussed in Section 5.4.

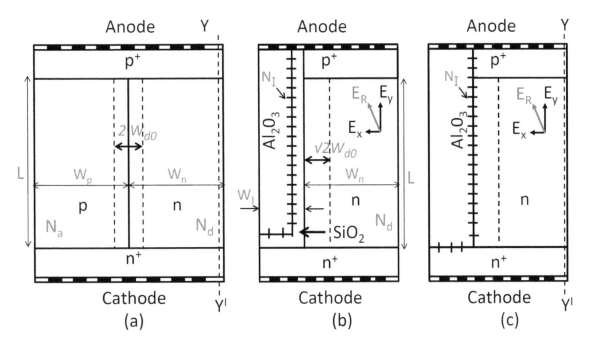

Fig. 5.1 Cross section of a *linear* cell of a SuperJunction (SJ) (a) and proposed Charge Sheet SuperJunction (CSSJ) (b). The device consists of lateral repetition of this cell. Diagram not to scale; the actual SO$_2$ liner is much thinner than shown. (c) A variation of (b) with SiO$_2$ liner removed and the negative interface charge N_I moved to the Al$_2$O$_3$/SiC interface.

5.1 DEVICE STRUCTURE AND SIMULATION SETUP

The proposed device is given in Fig. 5.1(b). The negative fixed charge N_I at the insulator / semiconductor interface can be varied in the range of 2.5–7.9 $\times 10^{12}$ cm^{-2} by varying the Al$_2$O$_3$ deposition temperature [6]. This negative charge sheet mimics the role of p- pillars of SJ and hence the name Charge Sheet Superjunction (CSSJ). A balanced CSSJ yields the least R_{ONSP} for a $V_{BR,target}$, and is designed to have $N_I = N_d \times W_n$. The structures considered in this work are compatible with the state of the art as per which the aspect ratio of the insulator (which fills trenches) and pillars can be as high as 18 [7].

The simulations use the Silvaco TCAD tool [8], Selberherr's impact ionization model for SiC with

$$a_n, a_p = 7.26, 6.86 \times 10^6 \, \text{cm}^{-1} \quad b_n, b_p = 23.4, 14.1 \, \text{MV cm}^{-1} \qquad (5.1)$$

calibrated against the data in [9], and a mobility, μ_n, dependent on the doping, N_d, of the n-pillar as per

$$\mu_n = 40 + \frac{950 - 40}{1 + \left(\dfrac{N_d}{2 \times 10^{17}}\right)^{0.76}} \quad cm^2\ V^{-1}\ s^{-1} \tag{5.2}$$

which fits into the measured data at $T = 300$ K [10].

The Silvaco device simulator employed by us can simulate the Al_2O_3 / SiO_2 / SiC system of Fig. 5.1(b) but only *without* the interface charge, N_I. This is because N_I happens to be at Al_2O_3 / SiO_2 which is an insulator / insulator interface, and the simulator does not allow placing of a charge at an insulator / insulator interface. Hence our device structure cannot be simulated as it is. We overcome this limitation by ignoring the SiO_2 liner and place N_I at the Al_2O_3 / SiC interface (see Fig. 5.1(c)) as the simulator allows the placing of an interface charge at a semiconductor / insulator interface. The electrical characteristics of the simulated device given by Fig. 5.1(c) and the actual device given by Fig. 5.1(b) has to be the same in both on- and off-state for this assumption to be justified. This has been done in the respective discussion of on- and off- state operation in the next section.

5.2 DEVICE OPERATION

The on- and off- state operation of CSSJ in SiC material is presented in comparison to that of SJ to highlight the advantages of the former. Most qualitative aspects presented here follow those reported in [1],[2] considering a Si device. However, the quantitative results are significantly different since the impact ionization and mobility parameters of SiC differ vastly from those of Si. Moreover, the R_{ONSP} formulae given below include the n-pillar depletion width, W_d, under zero-bias, denoted W_{d0}, which was ignored in prior works [1],[2].

5.2.1 On-State

Consider the cross section of an SJ (see Fig. 5.1(a)). Assuming $W_p = W_n$ without any loss of generality, the R_{ONSP} is given by

$$R_{ONSP} = \frac{2L}{q N_d \mu_n \left[1 - \left(W_{d0}/W_n\right)\right]}. \tag{5.3}$$

Here, q is the electron charge and μ_n is the electron mobility. Further, W_{d0} supports half of the built-in voltage V_{bi} of the junction between p and n pillars, and is given by

$$W_{d0} \approx \sqrt{\varepsilon_s V_{bi}/qN_d}, \quad V_{bi} \approx 2V_t \ln(N_d/n_i), \tag{5.4}$$

where ε_s is the dielectric constant, V_t is the thermal voltage and n_i is the intrinsic concentration. If the R_{ONSP} is sought to be reduced by reducing W_n and increasing N_d to maintain the charge balance, the V_{BR} degrades due to the increased peak field at the horizontal n$^+$/p-pillar interface. Instead, the CSSJ replaces the p-pillar of the SJ by an insulator film of thickness $W_I \leq W_n$ (see Fig. 5.1(b)). The negative interface charge, N_I, inverts the n-pillar inducing a vertical p$^+$n junction over L. The zero bias depletion width of this induced junction is $\sqrt{2}\,W_{d0}$ since it drops a potential $\approx V_{bi}$. The R_{ONSP} of CSSJ is given by

$$R_{ONSP} = \frac{L[1 + (W_I/W_n)]}{qN_d\mu_n[1 - (\sqrt{2}\,W_{d0}/W_n)]}. \tag{5.5}$$

Using the approximation $W_n \gg \sqrt{2}\,W_{d0}$ for simplicity, (5.3) and (5.5) show that, for a given N_d, the R_{ONSP} of the CSSJ is lower than that of SJ by the factor

$$\frac{R_{ONSP,CSSJ}}{R_{ONSP,SJ}} \approx 0.5\left(1 + \frac{W_I}{W_n}\right). \tag{5.6}$$

For all simulations, we use the device structure given in Fig. 5.1(c). The R_{ONSP} of the simulated device given by Fig. 5.1(c) and the actual device given by Fig. 5.1(b) can be argued and shown to be the same as follows. The parameters such as N_d, W_I and W_n that decides the R_{ONSP} as per (5.5) are chosen to be the same for the simulated device as that of the actual device. The equilibrium depletion width of the n-pillar, W_{d0}, is also the same for both the cases. This is because, in both cases the semiconductor interface with the insulator inverts forming a p$^+$ layer and the resultant depletion width is same as that in a p$^+$n junction diode. Hence, R_{ONSP} obtained by simulating the device in Fig. 5.1(c) is same as that of the actual device in Fig. 5.1(b).

Fig. 5.2 compares the N_d dependencies of the simulated R_{ONSP} of typical SJ and CSSJ. The devices simulated are balanced, but R_{ONSP} does not depend on the p-pillar doping, N_a. The CSSJ

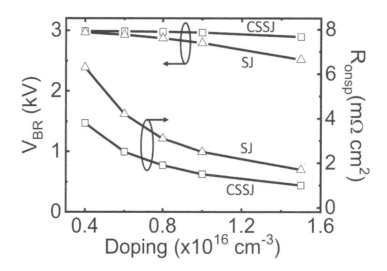

Fig. 5.2 Simulated dependence of the breakdown voltage and specific on-resistance on the n-pillar doping, N_d, of balanced SJ and CSSJ realized in 4H-SiC; $W_n = W_p = 5$ μm, $W_I = 1$ μm and $L = 18$ μm.

is seen to have 40% lower R_{ONSP} in accordance with (5.6). The difference in the N_d dependence of the V_{BR} of CSSJ and SJ, shown in this figure, is explained below.

5.2.2 Off-State

For the off-state study of a CSSJ, we estimate its avalanche breakdown voltage V_{BR} as a function of device parameters. The V_{BR} of the simulated device given by Fig. 1(c) will be same as that of the actual device given by Fig. 1(b), if (a) the breakdown occurs in SiC i.e. in the n-pillar rather than the insulator layers and (b) the electric field distribution in SiC remains the same for both the devices. These conditions can be argued to be true as follows. Firstly, it must be shown that the field in SiO$_2$ or Al$_2$O$_3$ is below the critical breakdown field of these insulators which is $\geq$ 5 MV/cm as against $\sim$ 3 MV / cm of SiC [11]. For this purpose, we use the simulated field and potential distributions at breakdown in a device with 20 % charge imbalance and $V_{BR} = 1$ kV, given in Fig. 5.3. This is the worst case scenario since the field is lower in devices with higher V_{BR} and lower charge imbalance.

We derive the accurate field distribution in all regions of the device by exploiting the fact that the SiO$_2$ liner thickness ($\sim$ 7 nm) is $<<$ Al$_2$O$_3$ (500 nm) or n-pillar thickness (700 nm). We obtain the field distribution in SiO$_2$ using Gauss law, as per which the field E_y *parallel* to the interface is continuous across the SiO$_2$ / SiC interface while the field E_x *normal* to the interface in SiO$_2$ is

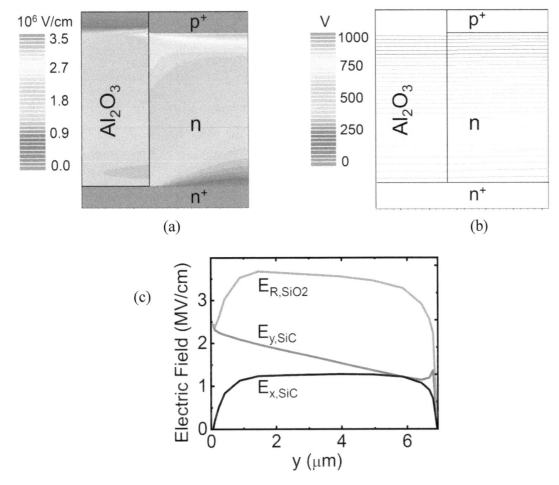

Fig. 5.3 Simulations at breakdown in a 4H-SiC CSSJ with W_n = 0.7 μm, W_I= 0.5 μm, L = 7 μm, N_d = 1 x 10^{17} cm^{-3}, $N_I = N_d W_n = 7 \times 10^{12}$ cm^{-2} and V_{BR} = 1 kV. (a) Field contours; (b) potential lines; (c) vertical and lateral components of the n-pillar field ($E_{y,SiC}$ and $E_{x,SiC}$), and the resultant field in SiO$_2$ liner ($E_{R,SiO2}$) along the Al$_2$O$_3$ / SiC interface over the pillar length, L.

$\varepsilon_{SiC}/\varepsilon_{SiO2}$ ≈ 2.5 times that in SiC; the resultant field in SiO$_2$ is therefore $E_{R,SiO2} = \sqrt{E_{y,SiC}^{2} + \left(2.5E_{x,SiC}\right)^{2}}$, where $E_{y,SiC}$ and $E_{x,SiC}$ are the simulated fields at the Al$_2$O$_3$ / SiC interface.

Fig. 5.3(a) confirms the following: the resultant field is maximum at the top right corner of the n-pillar, and hence maximum impact ionization or breakdown occurs at this point; the field reduces as one moves downward; the field in Al$_2$O$_3$ is less than that in SiC over the pillar length, and is everywhere < 1 MV / cm so that Al$_2$O$_3$ does not breakdown. The almost flat potential lines of Fig. 5.3(b) confirm that $E_y \gg E_x$ in SiC and Al$_2$O$_3$. Fig. 5.3(c) shows the distributions of

$E_{y,SiC}$, $E_{x,SiC}$ and $E_{R,SiO2} = \sqrt{E_{y,SiC}^2 + (2.5 E_{x,SiC})^2}$ over the pillar length. It is seen that $E_{R,SiO2}$ remains well below 5 MV/cm and so breakdown does not occur in SiO_2.

The validity of our simulations is confirmed by the fact that the simulated $E_{x,SiC} \approx q N_I / \varepsilon_{SiC} \varepsilon_0$ where $N_I = 7 \times 10^{12}$ cm^{-2}, over most of the pillar length. Further, this also justifies the assumption that the electric field distribution in SiC remains the same for the device in Fig. 1(c) and the actual device in Fig. 1(b).

5.2.2.1 Breakdown voltage of balanced devices

Fig. 5.2 shows that the V_{BR} of a balanced CSSJ is same as that of a comparable SJ at low N_d. However, CSSJ's V_{BR} does not degrade as N_d is raised unlike the SJ's V_{BR} for the following reason. The reverse bias applied across the CSSJ terminals transfers to the induced inversion layer / n-pillar p$^+$n junction, expanding its lateral depletion width, W_d. However, the body effect shrinks the inversion layer; the shrinkage is more at the n$^+$ end than the p$^+$ end due to 2-D effects. The formation and non-uniform shrinkage of the inversion layer as a function of reverse bias were discussed, illustrated pictorially and validated with TCAD in section IV (A) of our prior work on Si CSSJ [2]. The qualitative features of this work apply to SiC CSSJ as well.

As in zero bias, at other reverse biases too, W_d of CSSJ is $\sqrt{2}$ times that of SJ near the p$^+$ end where inversion layer is present. This is illustrated in Fig. 5.4(a) with the help of simulated potential lines in comparable SJ and CSSJ compatible with the state of the art [6], [7]. The nearly flat potential lines over W_I point to negligible voltage drop across the insulator thickness. Consequently, in CSSJ, the n-pillar gets fully depleted laterally at nearly half of the reverse bias required in an SJ; so this bias in CSSJ remains $<< V_{BR}$ even when N_d is raised. Once the n-pillar is fully depleted laterally, the field lines due to any more reverse bias emanating from the bottom n$^+$ directly terminate on the top p$^+$ as they find no charge left in the pillar to terminate on. Lower the bias for such lateral depletion, more the vertical field lines terminating directly or more uniform the vertical field distribution, E_y, over L, at breakdown. In corollary, even as N_d is increased, the breakdown field distribution in a CSSJ remains uniform unlike in a SJ (see Fig. 5.4(b)), and the area under this distribution, i.e. V_{BR}, remains constant.

It is of interest to compare the V_{BR} versus N_d behavior of SiC CSSJ (see Fig. 5.2) with that

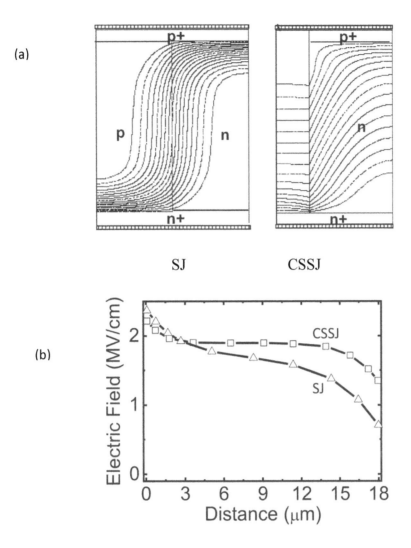

Fig. 5.4 (a) Simulated potential lines at 300 V illustrating early lateral depletion in CSSJ than SJ. (b) Simulated field distribution over the pillar length, L, at breakdown, along the cut-line YY$^|$ of Fig. 5.1. The devices are realized in 4H-SiC; $W_p = W_n = 5$ μm, $W_I = 1$ μm, $L = 18$ μm, $N_d = 1.5 \times 10^{16}$ cm^{-3} and $N_I = N_d W_n = 7.5 \times 10^{12}$ cm^{-2}.

of its Si counterpart with comparable geometry and same N_d range of 0.4 – 1.6 x 10^{16} cm^{-3} (see Fig. 4 of [1]). It is found that the V_{BR} of CSSJ in Si falls from 325 V to 150 V. The CSSJ's V_{BR} in both Si and SiC falls by ~ 175 V. However, the absolute value of V_{BR} in SiC is ~ 10 times higher than in Si due to the much higher E_C of the former. Hence, while the Si CSSJ's V_{BR} falls by a factor > 2, the relative fall in SiC CSSJ's V_{BR} is very small.

Since, at any N_d, CSSJ has a significantly lower R_{ONSP} and the same or slightly higher V_{BR} than an SJ, the CSSJ's R_{ONSP} is ~ 40% lower for a given V_{BR} (see Fig. 5.5).

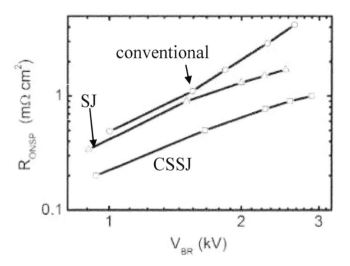

Fig. 5.5 Simulated specific on-resistance versus breakdown voltage for the SJ and CSSJ realized in 4H-SiC; $W_n = W_p = 5$ μm, $W_I = 1$ μm and $N = 1.5 \times 10^{16}$ cm^{-3}; L is varied to vary the V_{BR}. Conventional 4H-SiC junction data is shown for reference.

5.2.2.2 Breakdown voltage of imbalanced devices

The V_{BR} falls if the charge $N_d W_n$ in the n-pillar differs from the p-pillar charge $N_a W_p$ in an SJ or the interface charge, N_I, in a CSSJ. Such charge imbalance is inevitable due to process variations. For CSSJ, we define a charge imbalance factor as

$$k_{eff} = 1 - \left(N_I / N_d W_n \right) \quad N_I \le N_d W_n .$$ (5.7)

Here, we have considered the case $N_I \le N_d W_n$ rather than $N_I > N_d W_n$ since the V_{BR} is less for more n-pillar charge $N_d W_n$. We admit that k_{eff} is not a readily estimatable parameter in the form given in (5.7). Hence, it is of interest to express k_{eff} in terms of variables that are easily estimatable based on the knowledge of the technology employed in device fabrication. This would come handy for process and device design engineers. For this, let us assume that the target values of the device parameters N_d, W_n, and N_I are N_{d0}, W_{n0} and N_{I0} respectively. Here, we do not consider other parameters such as W_I and L as they do not affect k_{eff}. A balanced CSSJ yields the lowest R_{ONSP} for a $V_{BR,target}$, and hence we always design the parameters such that $N_{I0} = N_{d0} \times W_{n0}$. Using this, (5.7) can be rewritten as

$$k_{eff} = 1 - \left(\frac{N_I}{N_{I0}}\right)\left(\frac{N_{d0}}{N_d}\right)\left(\frac{W_{n0}}{W_n}\right) . \tag{5.8}$$

We define a doping imbalance factor,

$$k_N = 1 - \left(N_{d0}/N_d\right) \tag{5.9}$$

a width imbalance factor,

$$k_W = 1 - \left(W_{n0}/W_n\right) , \tag{5.10}$$

and a fixed charge imbalance factor,

$$k_{N_I} = 1 - \left(N_I/N_{I0}\right) . \tag{5.11}$$

Here, k_N can be estimated from the knowledge of the maximum doping variation of the epi-layer. Often, SiC wafers with epi-layers are procured for device fabrication and the information about the maximum lot to lot variation in doping concentration or resistivity is specified by the wafer manufacturing companies. k_W can be estimated from the knowledge of the tolerance of the lithography and etching technique used. k_{NI} may not be a familiar parameter for process engineers as this is unique to our device. Hence, we attempt to express it in terms of a more familiar parameter as follows.

The Al_2O_3 deposition temperature, T_{dep} can be expressed as a function of N_I by the following empirical fit to the measured data (see Fig. 5.6)

$$T_{dep} = 3.62 \times 10^{-11} N_I + 55.4 , \tag{5.12}$$

where T_{dep} is in °C and N_I is in cm^{-2}. Let T_{dep0} yield the value of N_{I0}. We define temperature error factor,

$$k_T = 1 - \left(T_{dep}/T_{dep0}\right)$$
$$= \Delta T_{dep}/T_{dep0} \tag{5.13}$$

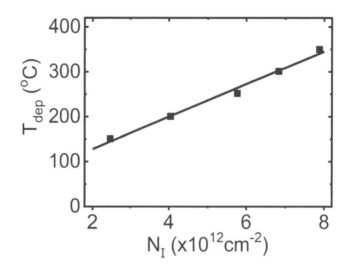

Fig. 5.6 The Al₂O₃ deposition temperature versus the negative charge concentration at the Al₂O₃/SiO₂ interface. Line is our model (5.24); points are experimental data from [6].

For process engineers, ΔT_{dep} and k_T are more readily available parameters compared to k_{NI}. Using (5.12) and (5.13) in (5.11), k_{NI} can be shown to be

$$k_{NI} = \left(1/k_T - 55.4/\Delta T_{dep}\right)^{-1} \tag{5.14}$$

The k_{eff} can be now re-written as

$$k_{eff} = 1 - \left(1 - k_{N_I}\right)\left(1 - k_N\right)\left(1 - k_W\right) \tag{5.15}$$

We can define the sensitivity of V_{BR} to k_{eff} as

$$S = \left[1 - \frac{V_{BR}(k_{eff})}{V_{BR}(k_{eff} = 0)}\right] \times 100 \tag{5.16}$$

Simulations have shown that the S of SJ and CSSJ realized in the same material differ by only < 5% [1]. However, it is of interest to compare the S of SJ or CSSJ realized in SiC with those realized in Si. Unlike the conventional approach in which devices having the same V_{BR} are compared, we follow a different approach here. We compare the S of SiC and Si devices having same R_{ONSP} and hence the same pillar parameters (assuming same μ_n). This comparison reveals a significant fact that the S of SJ or CSSJ realized in SiC is ~ 10 times lower than those realized in

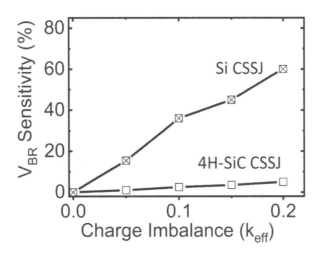

Fig. 5.7 Simulated V_{BR} Sensitivity versus charge imbalance of a Si and 4H-SiC CSSJ with $N_d = 7$ x 10^{15} cm^{-3}, $W_n = W_p = 2.5$ μm and $W_I = 1$ μm. For Si CSSJ, data is reproduced from [1].

Si. This is illustrated for the CSSJ in Fig. 5.7, where the simulated percentage degradation in V_{BR} is plotted against k_{eff}. Charge imbalance of significant degree may be present in a practical device, even with state of the art technology. Hence, apart from achieving a lower R_{ONSP} for a given V_{BR}, that has long been the motivation for replacing Si with 4H-SiC as per the literature, our approach suggests that the reduced sensitivity for a given R_{ONSP} is an added motivation to realize the CSSJ in 4H-SiC.

Considering the above advantages of CSSJ from physics point of view, we are motivated to examine the practicability of CSSJ fabrication in the next section.

5.3 DEVICE PRACTICABILITY

The potential steps for the fabrication of CSSJ has been discussed in the context of Si material in Ref. [2] and are summarized below. These steps can be adopted for fabricating CSSJ in SiC material.

1) Starting n$^+$ substrate

2) Epitaxial growth of the n-type drift layer

3) Aluminum implantation of the top p+ region

4) High aspect ratio trench formation using Inductively Coupled Plasma (ICP) etching

5) SiO$_2$ liner growth by dry oxidation

6) Al$_2$O$_3$ deposition by Atomic Layer Deposition (ALD) at a temperature, T_{dep}, in the range of 150−350 °C

7) Filling of the any unfilled trench by Chemical Vapor Deposited (CVD) Al$_2$O$_3$.

8) Contact formation

In the present work, we explain in detail the advantages of CSSJ from fabrication point of view by contrasting the above steps 4) – 7) used to realize the insulator with those used to realize the p-pillar of a SJ in 4H-SiC material.

The fabrication of SJ in 4H-SiC has been problematic. The first two reported works could not make a functional SJ device [12], [13]. The first functional 4H-SiC SJ was reported by Zhong et al. [14], [15] using trench etching and sidewall implantation of p-dopant. However, the fabrication of this 1.35 kV SJ device required six implantations of 40−360 keV energy followed by a high temperature anneal at 200−1400 °C for 30 minutes, for creating a p$^+$ liner along the sidewalls of a 6 μm deep trench. Also, the p-dopant activation efficiency varied between 20−65 % in the annealing temperature range, and required a trial and error approach to locate the temperature corresponding to the "optimum charge balance" that yields the maximum V_{BR}. This temperature was found to be 1350 °C and applies only to the device fabricated in Ref. [14]. A similar trial and error approach is required for fabricating devices with any other V_{BR}. Considering that the charge imbalance level in Si SJ can go up to 20 % [16], such imbalance levels could be even higher in 4H-SiC SJ due to this added difficulty in controlling the p-dopant activation efficiency [17]. In addition, the above steps of fabricating p-pillars in a 4H-SiC SJ are tedious, expensive and prone to causing severe wafer damage. We shall now assess the CSSJ steps 4) – 6) in the above backdrop.

Consider the trench width in the trench formation step 4). An SJ requires wider and hence lower aspect ratio trench to uniformly implant the trench side walls. The CSSJ could have a narrower trench than SJ, enabling reduction in R_{ONSP} as per (5.6). This is because, the CSSJ fabrication involves formation of trenches and subsequent insulator deposition on an SiO$_2$ liner, which is very well practiced in the case of trench dielectric isolation in CMOS and DRAM cells. These steps are also employed for creating isolation and edge termination in high voltage devices including superjunctions [18], [19]. In addition, highly conformal Al$_2$O$_3$ deposition by ALD could be achieved in trench with very high aspect ratio up to 50 [20].

Consider the trench depth in the trench formation step 4). Trenches of depth varying from few microns to > 100 μm may be required for making devices in the range of 1-10 kV or higher [21]. In SiC, high aspect ratio trenches with depth up to 53 μm have been fabricated using ICP etching [7] and up to 200 μm using laser ablation method [22]. However, for SJ devices, the number of high energy implantations, required to create a p⁺ layer with uniform charge along the walls, may significantly increase for deeper trenches [15]. This limits the voltage range of SJ devices realizable in 4-H SiC. However, the fabrication complexity of CSSJ does not increase with V_{BR} as multiple implantation steps are avoided, Thus, CSSJ is suitable for applications with a wider range of V_{BR}.

Consider the step 5) involving SiO_2 liner growth. The liner thickness is ~ 7 nm [6] but not critical for three reasons. First, it makes negligible contribution to W_I (see Fig. 5.1(b)) which happens to be > 100 nm due to the limitations of the trench realization process. Second, it does not affect N_I formed at the liner's interface with Al_2O_3 [6]. Third, there is negligible potential drop over this thickness even at high reverse bias as illustrated by the potential lines of the CSSJ in Fig. 5.4(a).

Consider the step 6) involving Al_2O_3 deposition by ALD to realize N_I at Al_2O_3 / SiO_2 interface. The magnitude of N_I can be controllably varied in the range

$$2.5 \times 10^{12} \leq N_I \leq 7.9 \times 10^{12} \text{ cm}^{-2}$$

(5.17)

by varying the deposition temperature, T_{dep}, in the range 150–350 ºC (see Fig. 5.6) [6]. The controllability of the charge at the SiO_2 / 4H-SiC interface does not affect the CSSJ operation. This is because, decades of research has reduced this charge from 5×10^{12} cm^{-2} [23] to 3×10^{11} cm^{-2} [24], i.e. to < 10 % of N_I of (5.17). Efforts are on to further reduce this charge which degrades inversion layer mobility of SiC MOSFETs. We recognize that the data of (5.17) and Fig 5.6 correspond to a Al_2O_3 / SiO_2 bilayer realized on a Si substrate [6]. Ideally, using this data for SiC substrate would require experimental verification which is beyond the scope of this work. Hence, we assume that this data can be applied for SiC substrate; the following reasons strengthen the validity of our assumption. Theories of N_I at the Al_2O_3/SiO_2 interface attribute N_I

to either the uncompensated negative $(AlO_{4/2})^-$ units at the interface [25], or the OH groups trapped in the volume of the Al_2O_3 [26]. In either theories, the nature and magnitude of N_I depend primarily on the Al_2O_3 and SiO_2 layers and their interface, and the role of the substrate on which these layers are deposited is secondary. Thus, [27],[28] report a negative charge of the same order as in (5.17) at the Al_2O_3/SiO_2 interface formed on 4H-SiC substrate and explain its origin using the theory of Ref. [6] based on Si substrates. This is analogous to the situation for the 2-Dimensional Electron Gas (2-DEG) at the AlGaN / GaN interface. The guidelines for the 2-DEG in AlGaN / GaN layers on sapphire substrate [29] are widely used for such bilayers on Si or SiC substrates as well.

Thermal cycles at 500−850 °C occur after the ALD of Al_2O_3 during possible trench filling by CVD deposited insulator [30] in step 7) or silicidation during contact formation [14], [31] in step 8). These cycles can be designed to affect the N_I minimally. For example, Ref. [6] showed that annealing at 700 °C for 60 minute causes only $< 5\%$ increase in N_I from the as-deposited value. Even if N_I turns out to be different than Ref. [6] for any reason or during further studies, our models and design procedure can still be used with suitable modification of the numbers in (5.17) and (5.12).

The CSSJ fabrication involves only a single high energy implantation process given in step 3), thereby saving cost and time, and reduces crystal damage and defects. Unlike this, for an SJ, multiple implantations of appropriately designed energy, dose and tilt angle are crucial in realizing a thin p$^+$ region with uniform charge along the walls of the trench. The large number of parameters involved makes this a tedious design problem. Process variations in all these parameters contribute to charge imbalance. In contrast, design of the p-pillar equivalent charge, N_I, in CSSJ is simpler as it involves choosing a single parameters, T_{dep}. Also, the easier control over the T_{dep} than the p-dopant implantation and activation parameters in SiC, enables fabrication of CSSJ with a lower charge imbalance than SJ.

Ref. [18] has given evidence for the lateral depletion of a SJ pillar by fixed charge of the insulator employed between p- and n-pillars to prevent dopant inter-diffusion. Ref. [2] has argued that insulator charges do not always pose reliability problems. For instance, in AlGaN / GaN HEMTs, a high interface charge (due to polarization) has been exploited to improve the device performance. Moreover, reliability problems posed by the trapped charge in the gate

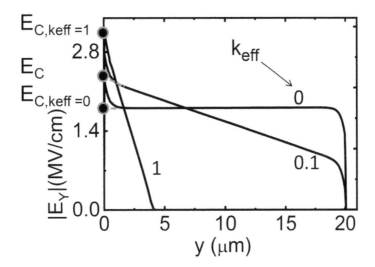

Fig. 5.8 Simulated vertical field distribution in a CSSJ at breakdown over pillar length along the cut-line YY' of Fig. 5.1 for different charge imbalance factors, k_{eff}. Pillar parameters are $L = 20$ μm, $W_n = 0.56$ μm, and $N_d = 8 \times 10^{16}$ cm^{-3}.

insulator of small-signal MOSFETs do not apply to the insulator charge providing the charge sheet essential for CSSJ operation. This is because the field in the insulator of a CSSJ is negligible (see Fig. 5.3(a)), unlike in the gate of a small-signal MOSFET.

Thus, CSSJ has the potential to solve the fabrication problems of SJ. We now discuss how the V_{BR} model and design procedure developed for SJ can be extended to CSSJ.

5.4 ANALYTICAL MODEL OF THE BREAKDOWN VOLTAGE

We derive a model for the V_{BR} in balanced and imbalanced CSSJ by considering the vertical breakdown field distribution, E_y, over L in a CSSJ along the cut-line Y-Y' in the n-pillar (see Fig. 5.1(b)). Fig. 5.8 shows this distribution for various $k_{eff} \leq 1$ as obtained from a TCAD simulator. We can regard this distribution as reducing linearly from an extrapolated peak value, E_C (the slope is zero for the balanced case $k_{eff} = 0$). This situation is analogous to that in an SJ analyzed in [21],[32], where the linear segment of the E_y versus y distribution has been shown to have the slope = $q k_{eff} N_d / 2\varepsilon_S$. Based on this approximation, we write

$$V_{BR} \approx E_C L - 0.25 \left(q k_{eff} N_d / \varepsilon_S \right) L^2$$

(5.18)

A simple formula for S defined in (5.16) can be derived using (5.18) and approximating E_C to be independent of k_{eff} as

$$S \approx 25qk_{eff}N_dL/\varepsilon_sE_C .$$

(5.19)

It is seen that $S \propto 1/E_C$ for a given value of N_d and L; as E_C of 4H-SiC is ~ 10 time higher than Si, the V_{BR} sensitivity of 4H-SiC CSSJ to charge imbalance is lower than that of Si CSSJ by the same factor.

For accurate estimation of V_{BR}, the dependence of E_C on L, N_d and k_{eff} should be taken into account. For this purpose, we adopt the following interpolating function given for an SJ [21]

$$E_C = (1 - k_{eff})^{\beta} E_{C,k_{eff}=0} + k_{eff}^{\beta} E_{C,k_{eff}=1}$$

(5.20)

where $E_{C,k_{eff}=0}$ is the value at low N_d and so a function of L alone, and $E_{C,k_{eff}=1}$ is the breakdown field in a 1-D p-n junction and so a function of N_d alone, as given below

$$E_{C,k_N=0} = (\gamma L)^{-1/8} , \quad E_{C,k_N=1} = \alpha N_d^{1/8}$$

$$\gamma = 4 \times 10^{-48} \text{ cm}^7.\text{V}^{-8} \qquad \alpha = 2.56 \times 10^4 \text{ V.cm}^{-5/8}$$

(5.21)

$$\beta \approx 0.8 + 1.85e^{-0.35r_n} \text{ for } r_n \geq 5.$$

Here, r_n denotes the aspect ratio of the n-pillar as per

$$r_n = L/2W_n .$$

(5.22)

The α and γ are based on the impact ionization coefficients of (5.1). The β expression has been verified upto $k_{eff} = 0.2$, and reduces to $\beta \approx 0.8$ for $r_n > 11$. Fig. 5.9 plots the E_C calculations using (5.20)-(5.22) versus V_{BR} calculated using this E_C in (5.18), and compares this plot with TCAD simulations, for $r_n \geq 11$ or $\beta = 0.8$. The agreement between the compared results validates the adoption of SJ's E_C formula for estimating CSSJ's V_{BR} using (5.18).

It is seen that the difference in the breakdown field distributions in the CSSJ and SJ at high N_d do not matter here. This difference is ultimately traced to the presence of inversion layer along the vertical interface of the n-pillar with the Al_2O_3 / SiO_2 in a CSSJ that is absent in the SJ. It is significant that the E_C and V_{BR} formulae developed for SJ apply to the CSSJ in spite of this

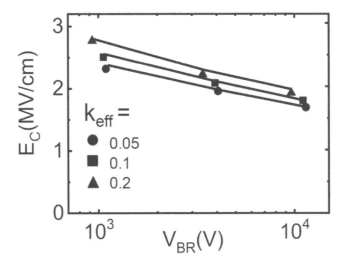

Fig. 5.9 Critical electric field as a function of V_{BR} for several values of charge imbalance factor, k_{eff}, and $r_n = 11$; values remain unaffected for $r_n > 11$. Lines are model results using (5.20)-(5.22) and (5.18) which apply to SJ [21]; solid circles are TCAD simulation results of CSSJ.

difference in the physics of the two devices.

5.5 DEVICE DESIGN

Recently [21], we gave an analytical procedure for designing a SJ having the minimum R_{ONSP} for a specified $V_{BR,target}$ and a k_{eff} governed by technology. We can adapt this procedure to design a CSSJ as follows. We set L and N_d of the CSSJ equal to the optimum n-pillar parameters – L_{opt} and N_{dopt} of the SJ. However, W_n of the CSSJ can differ from that of the SJ. Moreover, as pointed out in section III, the present SiC technology allows W_I of the CSSJ to be much thinner than the p-pillar of the SJ. We choose the maximum possible W_n and minimum possible W_I to minimize R_{ONSP} given by (5.5), i.e.

$$W_I = L_{opt}/2r_{I,\max}, \quad W_n = L_{opt}/2r_{n,\min}, \quad N_I = N_{dopt}W_{nopt}. \tag{5.23}$$

Here, $r_{I,max}$ is the maximum aspect ratio of the insulator permissible in the technology. Further, $r_{n,min}$ is the minimum aspect ratio of the n-pillar limited by the condition $N_I \leq N_{I,max}$ which is the maximum value in (5.17)

$$r_{n,\min} \geq \left(N_{dopt}L_{opt}/2N_{I,\max}\right). \tag{5.24}$$

Both r_I and r_n differ from aspect ratio r of the pillars of a SJ. Finally, we estimate R_{ONSP} from

(5.5), and the insulator deposition temperature using (5.12).

Ref. [21] gave the formulae for L_{opt} and N_{dopt} of an SJ as

$$L_{opt} \approx 2.08 \times 10^{-7} V_{BR\text{target}}{}^{8/7} \quad \text{for } r_n \geq 5, \ k_{eff} \geq 0.1,$$

(5.25)

whose values deviate from those of TCAD by $\leq 5\,\%$ in the validity range, and

$$N_{dopt} = \frac{2\varepsilon_s V_{BR\text{target}}}{q k_{eff} L_{opt}{}^2} \left[\frac{1 - \left(\sqrt{2}W_{d0}/2W_{nopt}\right)}{1 - \left(3\sqrt{2}W_{d0}/4W_{nopt}\right)} \right] \quad \text{for } k_{eff} \gg 1/2r_n{}^2,$$

(5.26)

where we have used the zero bias depletion width $\sqrt{2}W_{d0}$ of CSSJ instead of W_{d0} of SJ used in [21] and given by (5.4).

We can get a quick approximate estimate of the various parameters in closed-form using (5.23)-(5.26) in the following sequence: L_{opt} from (5.25), W_{nopt} and W_{Iopt} from (5.23), N_{dopt} from (5.26) where W_{d0} is calculated from (5.4) using the value of N_{dopt} with $W_{d0} = 0$, N_I from (5.23) and T_{dep} from (5.12).

For accurate results with a general k_{eff} and r_n, an iterative calculation is done as follows by including (5.18) and using L_{opt} from (5.25) as initial condition. A numerical calculator does this in < 1 s, while our MATLAB code [33] takes ≈ 70 ms.

a) $r_{n,min} = 1$, which is an initial guess.

b) $W_{nopt} = L_{opt}/2r_{n,min}$.

c) N_{dopt} from (5.26), where W_{d0} is estimated using N_{dopt} for $W_{d0} = 0$ for the first time, and using the previous N_{dopt} thereafter.

d) E_C from (5.20)-(5.22) and L_{opt} as the root of the quadratic (5.18). Iterate a)–c) until successive values of L_{opt} differ by $< 1\,\%$.

e) If $W_{nopt}N_{dopt} > N_{I,max} = 7.9 \times 10^{12}$ cm^{-2} increment $r_{n,min}$ by 0.2 and repeat a) – d).

f) $W_{Iopt} = L_{opt}/2r_{I,max}$

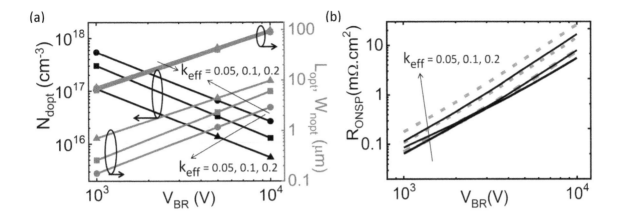

Fig. 5.10 (a) The n-pillar parameters of the CSSJ versus V_{BR} for different charge imbalance factors; the insulator thickness W_I is given by (5.23) using the L_{opt} data and $r_{I,max} = 18$ [7],[20]; $r_{n,min}$ is varied to yield the optimum results. Lines are our design results and points are TCAD simulations. (b) R_{ONSP} versus V_{BR} plots for CSSJ (solid lines) and for SJ with $r = 5$ (dashed lines)

Fig. 5.10 (a) shows that the above calculations of the CSSJ parameters for $V_{BR,target} = 1-10$ kV and $k_{eff} = 0.05 - 0.2$ agree with TCAD simulations. Fig. 5.10 (b) compares the R_{ONSP} of the CSSJ so designed with that of a SiC SJ having the same $V_{BR,target}$ and k_{eff}. It can be seen that for $V_{BR,target}$ in the range of $1-10$ kV, the R_{ONSP} of CSSJ is lower than that of SJ by 5–30%, 19–45%, and 36% for $k_{eff} = 0.05$, 0.1 and 0.2 respectively. Thus, CSSJ is a viable alternative to SJ in 4H-SiC material with superior electrical performance for high voltage 1-10 kV applications.

5.6 CONCLUSION

It was shown that the Charge Sheet SuperJunction (CSSJ) proposed earlier in Silicon (Si) has significant advantages over SJ in 4-H Silicon Carbide (SiC) material. These include, as compared to SJ in SiC, potentially simpler fabrication process, lower charge imbalance and 5–45 % lower specific ON-resistance for a given breakdown voltage. Further, a CSSJ in SiC is > 10 times less sensitive to charge imbalance than that in Si. The theory, modeling and design of SJ can be easily extended to CSSJ over 1-10 kV in spite of some differences in the physics of these two devices. Our work provides a strong motivation for fabricating the 4H-SiC CSSJ proposed.

5.7 REFERENCES

[1] S. Srikanth and S. Karmalkar, "Charge Sheet Superjunction (CSSJ): A new superjunction concept", in *Proc. IWPSD*, pp. 795–798, Dec. 2007.

[2] S. Srikanth and S. Karmalkar, "On the charge sheet superjunction (CSSJ) MOSFET", *IEEE Trans. Electron Devices,* vol. 55, no.12, pp. 3562-3568, Nov. 2008, doi: 10.1109/TED.2008.2006545.

[3] K. Akshay and S. Karmalkar, "Charge Sheet Super Junction in 4H-Silicon Carbide: Practicability, Modeling and Design," *IEEE J. Electron Devices Soc.*, vol. 8, pp. 1129 – 1137, Sep. 2020.

[4] K. Akshay and S. Karmalkar, "Note Clarifying "Charge Sheet Super Junction in 4H-Silicon Carbide: Practicability, Modeling and Design"," *IEEE J. Electron Devices Soc.*, vol. 8, pp.1315-1316, Oct. 2020.

[5] K. Akshay, M. G. Jaikumar and S. Karmalkar, "Charge Sheet Super Junction in 4H- Silicon Carbide," in *Proc. IEEE EDTM*, no. 19763722, Apr. 2020

[6] J. Buckley et al.,"Reduction of fixed charges in atomic layer deposited Al2O3 dielectrics", *Microelectron. Eng.*, vol. 80, no. 5, pp. 210-213, 2005, doi: https://doi.org/10.1016/j.mee.2005.04.070.

[7] K. M. Dowling, E. H. Ransom, and D. G. Senesky, "Profile evolution of high aspect ratio silicon carbide trenches by inductive coupled plasma etching", *J. Microelectromech. Syst.* vol. 26, no.1, pp. 135-142, Nov. 2017, doi: 10.1109/JMEMS.2016.2621131.

[8] ATLAS User's Manual, SILVACO International, 2006.

[9] A. O. Konstantinov, Q.Wahab, N. Nordell, and U. Lindefeltd, "Ionization rates and critical fields in 4H silicon carbide", *Appl. Phys. Lett.*, vol. 71, no. 1, pp. 90–92, Jul. 1997 doi: 10.1063/1.119478.

[10] M. Roschke and F. Schwierz, "Electron mobility models for 4H, 6H, and 3C SiC", *IEEE Trans. Electron Devices*, vol. 48, no. 7, pp. 1442–1447, Jul. 2001, doi: 10.1109/16.930664.

[11] G. D. Wilk, R. M. Wallace, and J. M. Anthony, "High-κ gate dielectrics: Current status and materials properties considerations", *J. Appl. Phys.*, vol. 89, no. 10, p. 5243, May. 2001, doi: https://doi.org/10.1063/1.1361065.

[12] R. Kosugi, Y. Sakuma, K. Kojima, S. Itoh, A. Nagata, T. Yatsuo, Y. Tanaka and H. Okumura, "Development of SiC super-junction (SJ) device by deep trench-filling epitaxial growth", *Mater. Sci. Forum*, vols. 740–742, pp. 785–788, 2013, doi: MSF.740-742.785.

[13] R. Kosugi, Y. Sakuma, K. Kojima, S. Itoh, A. Nagata, T. Yatsuo, Y. Tanaka and H. Okumura, "First experimental demonstration of SiC superjunction (SJ) structure by multi-epitaxial growth method", in *Proc. Int. Symp. Power Semiconductor Device ICs*, pp. 346–349, Jun. 2014, doi: MSF.778-780.845.

[14] X. Zhong, B. Wang, and K. Sheng, "Design and experimental demonstration of 1.35 kV SiC super junction Schottky diode", in *Proc. Int. Symp. Power Semiconductor Device ICs*, pp. 231–234, Jun. 2016, doi: 10.1109/ISPSD.2016.7520820.

[15] X. Zhong, B. Wang, J. Wang and K. Sheng, "Experimental Demonstration and Analysis of a 1.35-kV 0.92-mΩ.cm^2 SiC Superjunction Schottky Diode", *IEEE Trans. Electron Devices,* vol. 65, no.4, pp.1458-1465, 2018, doi: 10.1109/TED.2018.2809475.

[16] E. Napoli, H. Wang, and F. Udrea, "The effect of charge imbalance on superjunction power devices: An exact analytical solution", *IEEE Electron Device Lett.*, vol. 29, no. 3, pp. 249–251, Mar. 2008, doi: 10.1109/LED.2007. 915375.

[17] L. Yu and K. Sheng. "Modeling and optimal device design for 4H-SiC super-junction devices", *IEEE Trans. Electron Devices*, vol. 55, no. 8, pp. 1961-1969, Aug. 2008, doi: 10.1109/TED.2008.926648.

[18] S. Balaji and S. Karmalkar, "Effects of oxide-fixed charge on the breakdown voltage of superjunction devices", *IEEE Electron Device Lett.*, vol. 28, no. 3, pp. 229–231, Mar. 2007.

[19] K. P. Gan, X. Yang, Y. C. Liang, G. S. Samudra, and L. Yong, "A simple technology for superjunction device fabrication: Polyflanked VDMOSFET", *IEEE Electron Device Lett.*, vol. 23, no. 10, pp. 627–629, Oct. 2002, doi: 10.1109/LED.2002.803770.

[20] E. Shkondin et al., " Fabrication of high aspect ratio TiO_2 and Al_2O_3 nanogratings by atomic layer deposition ", *J. Vac. Sci. Technol. A*, vol. 34, no. 3, p. 031605, Apr. 2016, doi: 10.1116/1.4947586.

[21] K. Akshay and S. Karmalkar, "Quick Design of a Superjunction Considering Charge Imbalance Due to Process Variations", *IEEE Trans. Electron Devices*, vol. 67, no. 8, pp. 3024-3029, Aug. 2020.

[22] V. Khuat, Y. Ma, J. Si, T. Chen, F. Chen, and X. Hou, "Fabrication of through holes in silicon carbide using femtosecond laser irradiation and acid etching", *Appl. Surf. Sci.*, vol. 289, pp. 529–532, Jan. 2014, doi: 10.1016/j.apsusc.2013.11.030.

[23] L. A. Lipkin and J. W. Palmour, " Improved oxidation procedures for reduced SiO_2 /SiC defects ", *J. Electron. Mater.*, vol. 25, no. 5, pp. 909-915, May. 1996, doi: 10.1007/BF02666657.

[24] X. Yang, B. Lee, and V. Misra, "Electrical Characteristics of SiO_2 Deposited by Atomic Layer Deposition on 4H–SiC After Nitrous Oxide Anneal", *IEEE Trans. Electron Devices*, vol. 63, no. 7, pp. 2826-2830, May. 2016, doi: 10.1109/TED.2016.2565665.

[25] G. Lucovsky, J.C. Phillips, "Limitations for aggressively scaled CMOS Si devices due to bond coordination constraints and reduced band offset energies at Si-high-k dielectric interfaces", *Appl. Surf. Sci.*, vol. 166, no. 1-4, pp. 497-503, Oct. 2000, doi: 10.1016/S0169-4332(00)00482-7.

[26] P. Ericsson, S. Bengtsson, J. Skarp, "Properties of Al_2O_3-films deposited on silicon by atomic layer epitaxy", *Microelec. Engin.*, vol. 36, no. 1-4, pp. 91-94, Jun. 1997, doi: 10.1016/S0167-9317(97)00022-1.

[27] Usman M, PhD Thesis, Royal Institute of Technology KTH, 2012.

[28] Shukla, Madhup, G. Dutta, M. Ramanjaneyulu, and N. DasGupta. "Electrical properties of reactive-ion-sputtered Al_2O_3 on 4H-SiC", *Thin Solid Films*, vol. 607, pp.1-6, May. 2016.

[29] O. Ambacher, "Two-dimensional electron gases induced by spontaneous and piezoelectric polarization charges in N- and Ga-face AlGaN/GaN heterostructures", *J. Appl. Phys.*, vol. 85, pp. 3222-3233, Mar. 1999.

[30] A. N. Gleizes, C. Vahlas, M. M. Sovar, D. Samélor, and M. C. Lafont, "CVD-Fabricated Aluminum Oxide Coatings from Aluminum tri-iso-propoxide: Correlation Between Processing Conditions and Composition", Chem. Vap. Deposition , vol. 13, no. 1, pp. 23-29, Jan. 2007.

[31] S. K. Lee, C. M. Zetterling, M. Ostling, "Low resistivity ohmic contacts on 4H-silicon carbide for high power and high temperature device applications", *Microelectron Eng*, vol. 60, pp. 261-268, Jan. 2002, doi: 10.1016/S0167-9317(01)00603-7.

[32] M. Alam, D. T. Morisette, and J. A. Cooper, "Design guidelines for superjunction devices in the presence of charge imbalance", *IEEE Trans. Electron Devices,* vol. 65, no. 8, pp. 3345-3351, Aug. 2018, doi: 10.1109/TED.2018.2848584.

[33] https://codeocean.com/capsule/2622534.

CHAPTER 6

SUMMARY AND CONCLUSIONS

Superjunction (SJ) is a superior drift layer architecture that offers a lower R_{ONSP} than a 1-dimensional junction for a given V_{BR}. Analytical design of SJ is an important research problem due to the following reasons. First, it quickly gives deep physical insights into the optimal device design that are often difficult to obtain from a tedious trial and error based simulation study. Second, it can augment the data-driven approaches as follows. Data-driven approaches lack physical insights and consequently could lead to misinterpretation in the presence of poor quality data. As analytical designs are derived by considering physical effects, their accuracy is independent of the quality of the data at hand and hence the insights derived from it can help to avoid possible misinterpretations. Third, analytical solutions can be expressed in a material independent form which makes it useful for the design of SJ for a wide range of materials.

Chapter 3 derived material independent simple closed-form solutions for the optimum pillar doping, length and width of a balanced SJ. These parameters yield the minimum specific on-resistance, R_{ONSP}, for a target breakdown voltage, V_{BR}. The solutions have a generic form which allows quick calculations of R_{ONSP} versus V_{BR} of SJs for any given material. They were validated with TCAD simulations. The solutions reveal that optimum length depends on V_{BR} alone, and optimum doping on optimum width alone. Prior estimates of optimum doping were found to be lower and hence the R_{ONSP} higher than actual by > 30%. Thus, an improved theoretical minimum of the R_{ONSP} of a SJ has been established. Calculations of R_{ONSP} versus V_{BR} based on our solution were presented for materials ranging from Si whose band gap is low to SiC, GaN and diamond whose band gaps are high. These calculations brought out the scope for minimizing the R_{ONSP} of fabricated Si and SiC SJs (reported in literature) by process improvement. This should motivate process development for making SJs with lower R_{ONSP}.

Presently, it is believed that R_{ONSP} can be progressively reduced by raising r, and so literature has attempted pillar aspect ratios > 25 (10) in Si (4H-SiC) SJs. Counterintuitively, we showed in Chapter 4 that, in practical SJs with charge imbalance > 5 %, the R_{ONSP} attains a minimum at an optimum $r = r_0$ which is as low as 8–15 for Si and 4H-SiC SJs with $V_{BR} = 0.1$–10 kV; the r_0 is not sharply defined, as even $\pm$ 30% change in r around r_0 raises R_{ONSP} by <10% above the minimum. Moreover, the r_0 increases logarithmically with (V_{BR}/Bandgap) for $k \geq 2.5/r^2$; in contrast, the r_0 of a balanced SJ increases super-linearly with (V_{BR}/Bandgap) and is several tens of times higher, calling for caution while using balanced SJ theory to design practical SJs. We gave simple closed-form SJ design equations valid across materials based on the new insight. These equations show that as compared to an optimum balanced device, a device optimized considering charge imbalance has significantly lower pillar doping, higher pillar width, but almost same pillar length. Further, we showed that our solution can be used to obtain SJs having lower R_{ONSP} than prior works for the same V_{BR} and yet a much lower and comfortable r that is easier to fabricate.

Chapter 5 showed that the Charge Sheet SJ (CSSJ) proposed earlier in Silicon (Si) has significant advantages over SJ in 4-H Silicon Carbide (SiC) material. These include, as compared to SJ in SiC, potentially simpler fabrication process, lower charge imbalance and 5–45 % lower R_{ONSP} for a given V_{BR}. Further, a CSSJ in SiC is > 10 times less sensitive to charge imbalance than that in Si. The theory, modeling and design of SJ can be easily extended to CSSJ over 1-10 kV in spite of some differences in the physics of these two devices. Our work provides a strong motivation for fabricating the 4H-SiC CSSJ proposed.

Milton Keynes UK
Ingram Content Group UK Ltd.
UKHW050620080324
438959UK00012B/462